# ─── *Author's Acknowledgements* ───

I would like to express my sincere thanks to the friends at my workshop for editing, trailing, advising and reporting on the materials.

*S K So*
*Chief Author*

The Publisher welcomes any comments from readers. Please email us at sosir2006@hotmail.com. For school order, please contact the Marketing Department on (852) 9034 9915.

# LCCI初級及中級
# 會計天書 ✔ DISTINCTION
# 必勝攻略【第二版】

| | |
|---|---|
| 作者 | 蘇斯杰 ( So Sze Kit ) , PhD ( Cand. ) , MBA ( OUHK ) , PGDE ( CUHK ) |
| 出版 | 超記出版社（超媒體出版有限公司） |
| 地址 | 荃灣柴灣角街 34-36 號萬達來工業中心 21 樓 02 室 |
| 電話 | ( 852 ) 3596 4296 |
| 電郵 | info@easy-publish.org |
| 網址 | http : //www.easy-publish.org |
| 香港總經銷 | 聯合新零售（香港）有限公司 |
| 上架建議 | 商業會計 |
| ISBN | 978-988-8839-20-9 |
| 定價 | HK$98 |

U0152119

Printed and Published in Hong Kong
版權所有 · 侵害必究

# Level 1

## 《LCCI 初級會計(Level 1) 精讀備試天書》

 ontent 目錄 / / / /

本書附送 LCCI 增值錦囊，請到以下網址下載：
http://www.systech.hk/lcci-note.rar

# *Level 2*

## 《LCCI Level 2 中級會計精讀備試天書》

# Content 目錄 ////

本書附送 LCCI 增值錦囊，請到以下網址下載：
http://www.systech.hk/lcci-note.rar

# 會計這一科真的值得讀嗎？

## 為何要修讀 LCCI 會計課程

　　LCCI 為 London Chamber of Commerce & Industry 的簡稱，已經成立了超過一百多年。LCCI 考試提供 60 種以上的專業考試，而其中會計是最為香港人所熟悉的。LCCI 前主席 Sir James Duncan 曾表示，得到 LCCI 會計資格，猶如得到 Passport to Employment（就業通行証）。如留意香港報章的求職版，最多的職位空缺是業務代表，其次就是會計、財務、金融類等行業，所以擁有 LCCI 會計證書是非常重要的。

　　另一方面，從投資角度看，考取 LCCI 會計證書亦是一個合適的投資。以香港職場的文員為例，持有 LCCI 會計證書除了入職會較優先外，起薪點亦平均較高，約 300-500 元。

考取 LCCI 所需的開支：

| | |
|---|---|
| 修讀會計課程 | $1000 |
| 會計教科書（多本） | $200 |
| 考試費 | $400 |
| 總投資數 | $1,600 |

　　如果入職多了 $300，那麼，5 個月便回本了，第六個月便是賺。比投資開店做生意還要快，作為精明的你，又怎會放過這機會？

報考方法

| 季度 | 報考日期 | 考試日期 |
|---|---|---|
| Series 2 | Oct－Nov | 下年 Apr－May |
| Series 3 | Dec－Jan | 下年 Jun－July |
| Series 4 | Jun－Jul | 同年 Nov－Dec |

・報表格可向就讀學校提出

・可向 LCCI 香港辦事報告，查詢電話：(852) 3102 0100

・如有疑問，也可向作者蘇老師查詢：sosir2006@hotmail.com

報考須知

・每季考試均有兩星個日期，考生無權選擇應考日期，而日後不能改的，考生必須準備讓兩天時間。

・准考證會於開考約前四星期送到報名學校，考生請記緊自行致電報名學校查詢。

・收到准考證後留意姓名、身份證號碼是否正確。

# LCCI 考試之準備

◎ 溫習重點

LCCI Level 1 重點：
- Bank Reconciliation Statement
- Petty Cash Book and the Imprest System.
- Capital and revenue expenditure.
- Adjusting for accruals and prepayments.
- Depreciation of fixed assets.
- Bad debts and provision for doubtful debts.
- T, P&L a/c 及 Balance Sheet

LCCI Level 2 重點：
- Partnership
- Limited Company
- Single Entry and Incomplete Record
- Trading, Profit & Loss Account
- Balance Sheet
- Income & Expenditure a/c

◎ 時間分配

- 考生可用大約 35 分鐘去完成一題。
- 不要放棄作答任何一條試題，要盡量分配時間作答所有題目。

◎ 100 天前準事宜

- 計數機印
- 考生應至少溫習三年以上的歷屆試題，才能真正掌握出題趨勢及作答要求。
- 溫習 Past Paper 時，不能只看答案，記著要白紙黑字做出來呢！
- 溫習如有不明白之處，立即向作者蘇老師提問：sosir2006@hotmail.com
- 向公司申請考試當日請假半天。
- 事前要探路，了解如何到達試場，除了的士外，還有何途徑。

◎ 試前一晚的準備

1. 文件
   - 准考證 (仔細檢查檢查日期)
   - 身份證
2. 文具安排
   - 鉛字筆 (2 或更多)
   - 透明間呎 (2 或更多)
   - 計數機 (已蓋印)
   - 改錯帶
   - 熒光 / 顏色筆
   - 鉛筆
3. 運輸安排
   - 前往路線 (多於一個的選擇)

　　　　・硬幣

　　　　・八達通

4. 食品安排

　　　　・晚餐

　　　　・蒸餾水

5. 衣服

　　　　・外套

　　　　・運動鞋（方便「追」巴士）

　　　　・手巾 / 紙巾

　　　　・髮夾

6. 醫學安排

　　　　・Panadol

　　　　・香薰油或白花油（提神用的）

◎ 注意事項

・先花三數分鐘時間來審核試題

・選擇最有把握的題目作答，然後才作答較沒把握的題目。

・仔細閱讀 REQUIRED 的部分

・Underline 關鍵及重要字（利用不同顏色）

・所有作答均要清楚及整齊

・緊守 35 分鐘一條的時間控制策略，不能超時。

・嘗試回答所有問題

・一條都不能少

◎ 分數表現

・Pass → 50%

・Credit → 60%

・Distinction → 75%

◎ 評分準則

・答對地方便有分

・如果計算錯誤，但計算步驟清楚列明，有機會得到部分分數

◎ 後記

・充份準備才能使你的壓力下降！

　　　　如想有 Credit 或以上，只靠溫習課本是不足夠的，必須要大量操練試題。你試過操練時遇到問題不知如何是好嗎？本書作者從事會計培訓多年，深明問題之所在，為了協助勤力的學生，現特設網絡支援服務，包括：電郵問書服務、最近模擬試題、模擬試題批改等等。如果你對考試範圍有疑問，可將題目電郵至 sosir2006@hotmail.com 向作者發問，之後便會有專人回覆及跟進，為你解決疑難。

# LCCI 升學途徑

Level 1: Book-keeping

Level 2: Book-keeping & Accounts

Level 2: Cost Accounting

Level 2: Computerised Accounting

24個月內完成Level 2這三科，便可得到LCCI發出的 Group Cert.

Level 3: Accounting

Level 3: Cost Accounting

Level 3: Computerised Accounting

24個月內完成Level 2這三科，便可得到LCCI發出的 Group Dip.

Level 4: Financial Accounting

Hong Kong Institute of Accredited Accounting Technicians (HKIAAT)

Paper 1: Financial Accounting

如LCCI Level 2考試中，得到Credit或以上之成績，可申請豁免這一張卷。

# *Level 1*

《LCCI 初級會計 (Level 1) 精讀備試天書》

## Content 目錄 ////

本書附送 LCCI 增值錦囊，請到以下網址下載：
http://www.systech.hk/lcci-note.rar

# Chapter 1

## Introduction and Day Books

## Content

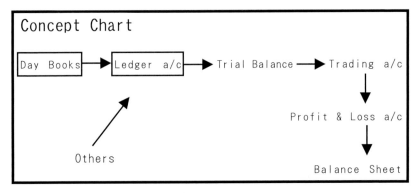

Concept Chart

Day Books → Ledger a/c → Trial Balance → Trading a/c

Others

Profit & Loss a/c

Balance Sheet

## [01] What is Accounting?

Accounting Concerned

Economics Information + Financial Information

Function

Better Decision + Budget for future

## [02] Process of Accounting

In general, the process of accounting can be divided into 5 stages:

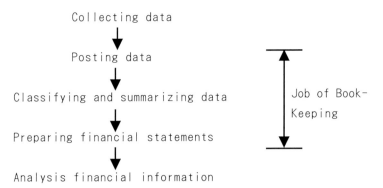

Collecting data

↓

Posting data

↓

Classifying and summarizing data

↓

Preparing financial statements

↓

Analysis financial information

Job of Book-Keeping

8

A professional accountant has the responsibility of carrying out all of the above stages.

## [03]   What is book-keeping?

◎ to measure accounting data of an undertaking
◎ to communicate the business results to the interested parties such as board of directors.

## [04]   Users of Accounting Information

### a.Owner(s) of the business

◎ They want to know the business is profitable or not.
◎ They want to know what the financial resources of the business are.

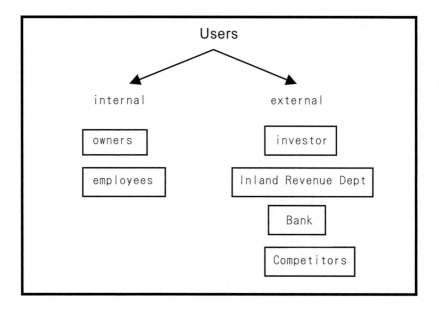

## b.A prospective investor 準投資者

When the owner wants to sell it, the potential investor must want to see such information.

## c.The bank

If the owner wants to borrow money for the business, then the bank will need such information.

## d.The Inland Revenue Department 税務局

They need it to calculate the taxes payable.

## e.Competitors

Competitors are interested to evaluate their own efficiency and to monitor competition.

## f.Employees

They concern on their job security and remuneration.

## [05]　Accounting Equation

The whole idea of financial accounting can be summed up by the accounting equation which is shown as follows:

After making profit:

Assets = Liabilities + Capital + Income − Expenses

Assets + Expenses = Liabilities + Capital + Income

10

## a. Assets (A)

They are economic resources, which help to make profit

Fixed Assets FA

◎ are assets to be kept as such for more than one year

e.g. Land and Buildings, Furniture & Fitting, Office Equipment

Current Assets CA

◎ are assets which change from day to day

e.g. Stock, Debtors, Bank, Cash

## b. Liabilities (L)

They are debts or obligations owed by the business to outside parties.

Long-term liabilities / Amount due more than one year LL

◎ are those liabilities which have to be paid more than one year

e.g. Bank loans, 5% Debentures

Current liabilities / Amount due within one year CL

◎ are those liabilities which have to be paid within one year.

e.g. Creditors, Bank overdrafts, Accrual Expenses

## c. Capital (C)

The total of resources supplied by owner. It is a liability owed to its owners.

## d. Income (I)

It can increase the capital.

e.g. Sales, Rent received, Interest received

## e. Expenses (E)

Expenses refer to all expenditures incurred in the usual courses in earning income.

(i) Trading Expenses e.g. Purchases, Carriage inwards

(ii) Administrative Expenses e.g. Rent and rates

(iii) Selling and Distribution Expenses e.g. Carriage outwards, Commission

(iv) Financial Expenses e.g. Bank interest

## Classwork 1

For each of the following items, put the symbol in the space provided to show the major accounting category.

Symbols:  Assets: A;  Capital: C;  Liability: L

1. Capital introduced_____ C _____
2. Premises _____
3. Stock _____
4. Loan from Chan _____
5. Trade debtors _____
6. Loan to May _____
7. Creditors _____
8. Motor vehicles _____
9. Deposit at HSBC Bank _____
10. Invest in shares of Tom Online _____

## Classwork 2

Supply the missing amount of the following cases:

| | Assets | Liabilities | Capital |
|---|---|---|---|
| | $ | $ | $ |
| a) | 50,000 | 19,000 | _____ |
| b) | _____ | 19,300 | 54,400 |
| c) | 38,000 | _____ | 28,000 |
| d) | 219,500 | 25,500 | _____ |
| e) | 78,000 | _____ | 40,000 |

## [06]　The Sales Journal and The Sales Ledger

### a.Sales

| Cash sales | Credit sales |
|---|---|
| When cash receives for selling goods immediately, there is <u>no</u> need to enter such sales <u>in</u> the sales journal. | In many businesses most of the sales will be made on credit |

### b.Sales Journal (Sales Day Book)

From the copies of the sales invoices, the book-keepers record data to their sales journals.

### c.Sales Ledger

It is used for recording customers' (debtors') record.

### d.The Sales Journal and the Sales Ledger

## Posting credit sales ONLY

## Classwork 3

Mr. Mak has the following sales for the month of March 20X7.

| | |
|---|---|
| Mar 1 | Sold goods on credit to Lily Ltd. for $40. |
| Mar 5 | Sold goods on credit to Lily Ltd. for $200 less 20% trade discount. |
| Mar 25 | Credit sale to Betty Ltd. for $100. |
| Mar 27 | Cash sales $100 to Betty Ltd |
| Mar 30 | Sold goods on credit to ABC Company for $40 less 10% trade discount. |

## Sales Journal / Sales Day Book

| Date | Particulars | F | Amount ($) | Total ($) |
|---|---|---|---|---|
| | | | | |
| | | | | |
| | | | | |
| | | | | |
| | | | | |
| | | | | |

## [07]　The Purchases Journal and the Purchases Ledger

### a.Purchases

(i)Cash purchases

When cash pays for goods immediately, there is _no_ need to enter such purchases in the purchases journal.

(ii)Credit purchases

In many businesses most of the purchases will be made on credit for immediate cash.

### b.The Purchases Journal

It is used for recording all credit purchases.

### c.The Purchases Ledger

It is used for recording suppliers' (creditors) record.

### d.The Purchases Journal and the Purchases Ledger

## Posting Credit Purchase ONLY

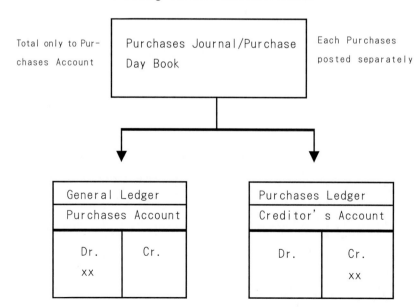

| Total only to Purchases Account | Purchases Journal/Purchase Day Book | Each Purchases posted separately |

| General Ledger | | Purchases Ledger | |
|---|---|---|---|
| Purchases Account | | Creditor's Account | |
| Dr. | Cr. | Dr. | Cr. |
| xx | | | xx |

# Classwork 4

Mr. Tang has the following transactions during the month of March 20X7 :

| | |
|---|---|
| Mar 4 | Purchased goods on credit from Yuen Ltd. for $1,400 less 5% trade discount. |
| Mar 15 | Purchased goods on credit from Yuen Ltd. for $1,800 for 10% trade discount. |
| Mar 26 | Purchased goods on credit from Main Main Ltd. for $3,100. |
| Mar 29 | Purchase goods from Main Main Ltd $2,600 by cheque |

## Purchase Journal / Purchase Day Book

| Date | Particulars | F | Amount ($) | Total ($) |
|------|-------------|---|------------|-----------|
|      |             |   |            |           |
|      |             |   |            |           |
|      |             |   |            |           |
|      |             |   |            |           |
|      |             |   |            |           |

## [08]　The Returns Inwards Journal

a. Returns Inwards and Credit Notes

(i) Returns inwards journal

It is used to record goods returned to us

(ii) Credit notes

It is a document sent to a customer by the business, showing allowance given by supplier in respect of unsatisfactory goods.

### Posting Return Inwards

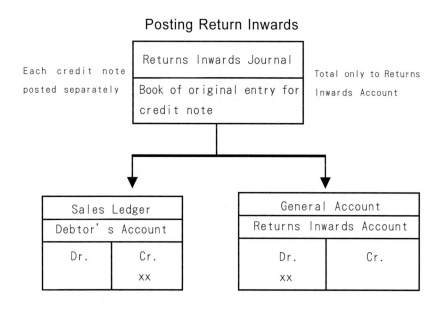

Each credit note posted separately

**Returns Inwards Journal**

Book of original entry for credit note

Total only to Returns Inwards Account

| Sales Ledger | |
|---|---|
| Debtor's Account | |
| Dr. | Cr. |
|  | xx |

| General Account | |
|---|---|
| Returns Inwards Account | |
| Dr. | Cr. |
| xx |  |

## Classwork 5

An example of a returns journal:

## Returns Inwards Journal

| Date | Particulars | Folio | Amount |
|------|-------------|-------|--------|
| 20X7 | | | $ |
| Sep 2 | Chun Store | SL 1 | 400 |
| 17 | Peter Ltd. | SL 2 | 120 |
| 19 | Amy | SL 3 | 290 |
| 29 | Super Store | SL 112 | 160 |
| 30 | Total transferred to Returns Inwards Account | GL 1 | 972 |

## Return Inwards

## Debtor

## [09] The Returns Outwards Journal

### a.Returns Outwards and Debit Notes

(i) Returns outwards journal

It is used to record goods returned to our suppliers.

(ii)Debit Notes

It is a document sent by a business to their supplier, giving details of a claim for an allowance in respect of unsatisfactory goods.

### Posting returns outwards

### Classwork 6

An example of a returns outwards journals:

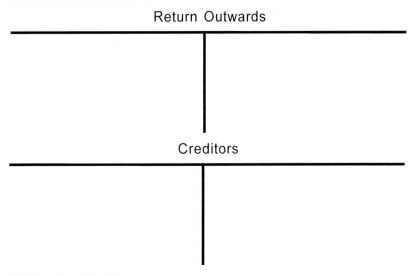

| Returns Inwards Journal | | | |
|---|---|---|---|
| Date | Particulars | Folio | Amount |
| 20X7 | | | $ |
| Sep 11 | John | PL 29 | 180 |
| 16 | X Ltd. | PL 46 | 100 |
| 28 | Silly | PL 55 | 30 |
| 30 | Jimmy | PL 4 | 360 |
| 30 | Total transferred to Returns Outwards Account | GL 2 | 690 |

## Return Outwards

## Creditors

## [10]  Examination

| 2003 | | S3 | |
|---|---|---|---|
| 2004 | | S3 | |
| 2005 | S2 | S3 | S4 |
| 2006 | S2 | | |

# Chapter 2

## Double Entry System

## Content

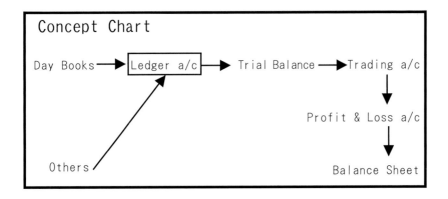

**Concept Chart**

Day Books → Ledger a/c → Trial Balance → Trading a/c → Profit & Loss a/c → Balance Sheet

Others → Ledger a/c

## [01]   Nature of a transaction

Events that result in such changes are known as transactions.

## [02]   Double entry system

Account is part of double entry records, containing details of transactions for a specific item.

Double entry bookkeeping is a system by which each transaction is entered twice.

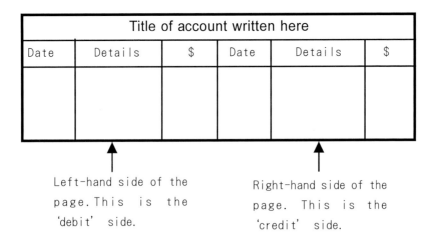

| Title of account written here | | | | | |
|------|---------|-----|------|---------|-----|
| Date | Details | $ | Date | Details | $ |
| | | | | | |

Left-hand side of the page. This is the 'debit' side.

Right-hand side of the page. This is the 'credit' side.

## [03] Double entry rules

|  | DR | CR |
|---|---|---|
| Income (I) | – | + |
| Expenses (E) | + | – |
| Assets (A) | + | – |
| Liabilities (L) | – | + |
| Capital (C) | – | + |

## [04] Stock movements

| Account | Reason |
|---|---|
| Purchases account | For the purchase of goods for resale |
| Sales account | For the sale of goods in the normal of business |
| Returns Inwards account | For goods returned to the firm by its customers |
| Returns Outwards account | For goods returned by the firm to its suppliers |

## [05] Purchases of stock on credit

Example 1 : 1 August 20X7.  Goods costing $1,000 were bought on credit from Super Store.

## [06]  Purchases of stock for cash

Example 2 : 2 August 2003.   Goods costing $200 was bought, cash being paid for them immediately.

## [07]  Sales of stock on credit

Example 3 : 3 August 2003.   Sold goods on credit for $250 to Lee.

## [08]  Sales of stock on cash

Example 4:   4 August 2003.   Goods were sold for $500, the cash being received at once upon sale.

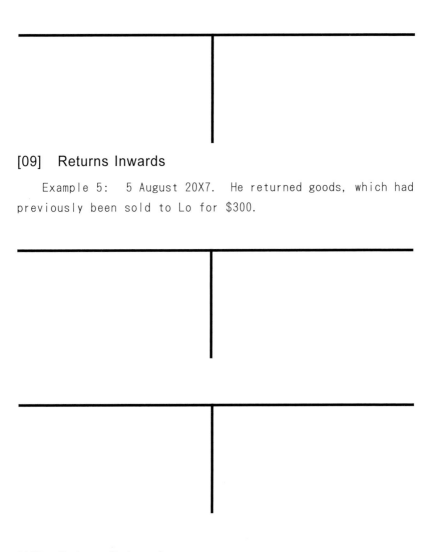

## [09] Returns Inwards

Example 5:　5 August 20X7.　He returned goods, which had previously been sold to Lo for $300.

## [10] Return Outwards

Example 6:　6 August 2003.　Goods previously bought for $900 were returned by the firm to K. So.

## [11]   Revenues

Example 7:   5 June 20X7. We let someone else use our premise and received rent of $40 by cheque.

## [12]   Drawings

Drawings consist of cash or goods taken out of a business by the owner for his private use.

Example 8:   25 August 2003.   Proprietor took $100 cash out of the business for his own use.

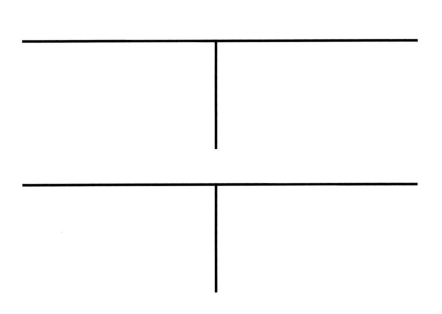

## [13]   Payment of expenses

Example 9:   26 Aug 2003, we paid $2,500 of rent by cheque.

## [14]   Assets Increase

Example 10:  Buy car $10,000 from ABC Ltd $10,000, we paid $1,500 for deposit, the balance will pay one month later.

## [15]    Capital Movement

### Capital

| | |
|---|---|
| Drawings | Bal b/d |
| Net Loss | Cash / Bank |
| Bal c/d | Net Profit |

## [16]    Balancing

It means locating the difference between two sides on the smaller side, so as to equalize the totals on each side of the account.

### Example 11

### Debtors

| 2005 | | $ | 2005 | | $ |
|---|---|---|---|---|---|
| Jan 1 | Sales | 1,000 | Jan 2 | Bank | 250 |
| Jan 3 | Sales | 450 | Jan 4 | Cash | 20 |
| | | | Jan 31 | Bal c/d | |

## [17]    Examination

| 2003 | | S3 | S4 |
|---|---|---|---|
| 2004 | S2 | | S4 |
| 2005 | S2 | S3 | |
| 2006 | S2 | | |

# Chapter 3

Trial Balance

## Content

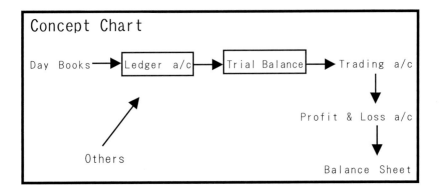

Concept Chart

Day Books → Ledger a/c → Trial Balance → Trading a/c
                                                    ↓
                                          Profit & Loss a/c
                                                    ↓
Others ↗                                   Balance Sheet

## [01]   Balancing off Accounts

At the end of the accounting period, all accounts should be balanced off.

The balance over the totals is called as balance carried forward (bal c/f) or balance carried down (bal c/d).   The balance below the total is called as balance brought forward (bal b/f) or balance brought down (bal b/d).

Example 1:

### Debtor - Mei

| 2004 | | $ | 2004 | | $ |
|---|---|---|---|---|---|
| Mar 3 | Sales | 158 | Mar 18 | Bank | 158 |
| Mar 15 | Sales | 206 | | | |
| Mar 28 | Sales | 118 | Mar 31 | Bal c/d | 324 |
| | | 482 | | | 482 |
| Apr 1 | Bal b/d | 324 | | | |

Example 2:

## Creditor - Lee

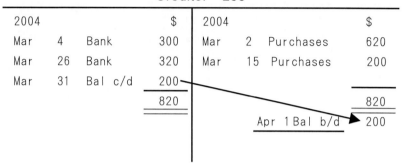

| 2004 | | | $ | 2004 | | | $ |
|------|---|------|-----|------|----|-----------|-----|
| Mar | 4 | Bank | 300 | Mar | 2 | Purchases | 620 |
| Mar | 26 | Bank | 320 | Mar | 15 | Purchases | 200 |
| Mar | 31 | Bal c/d | 200 | | | | |
| | | | 820 | | | | 820 |
| | | | | Apr | 1 | Bal b/d | 200 |

## [02]  Different cases of balance c/d

1. Cases of balancing the account $\boxed{Dr = Cr}$

### Peter

| 2006 | | | $ | 2006 | | | $ |
|------|---|-------|-----|------|----|---------|-----|
| Jan | 1 | Sales | 200 | Jan | 2 | Cash | 100 |
| | 4 | Sales | 500 | | 4 | Cash | 600 |
| | | | | | 31 | Bal c/d | |
| | | | 700 | | | | 700 |

2. Cases of balancing the account $\boxed{Dr > Cr}$

### Peter

| 2006 | | | $ | 2006 | | | $ |
|------|---|-------|-----|------|----|---------|-----|
| Jan | 1 | Sales | 200 | Jan | 2 | Cash | 100 |
| | 4 | Sales | 500 | | 4 | Cash | 450 |
| | | | | | 31 | Bal c/d | |
| | | | 700 | | | | 700 |

3. Cases of balancing the account $\boxed{Dr < Cr}$

## Peter

| 2006 | | | $ | 2006 | | | $ |
|------|---|-------|-----|------|---|---------|-----|
| Jan | 1 | Sales | 100 | Jan | 2 | Cash | 100 |
| | 4 | Sales | 500 | | 4 | Cash | 600 |
| | 31 | Balc/d | | | | | |
| | | | | | 31 | Bal c/d | |
| | | | 700 | | | | 700 |

## Classwork 1

| Jun | 1 | Cash sales $1,000 |
|-----|-----|--------------------|
| | 9 | Credit Sales $1250 |
| | 12 | Cash Purchase $200 |
| | 23 | Cash Sales $1,000 |
| | 30 | Paid Rent $50 |

Required:

Prepare the Cash a/c and balance off

## Cash

| 2006 | $ | 2006 | $ |
|------|---|------|---|
| Jun | | Jun | |

## [03]  Format of Trial Balance

At the end of the accounting period, all accounts should be balanced off or closed. All the balances are extracted to a trial balance.

# K. Poon

## Trial Balance as at 31 May 2001

|  | Dr. | Cr. |
|---|---|---|
|  | $ | $ |
| Bank | xxx |  |
| Cash | xxx |  |
| Yip |  | xxx |
| Mok | xxx |  |
| Capital |  | xxx |
| Purchases | xxx |  |
| Sales |  | xxx |
| Returns inwards | xxx |  |
| Returns outwards |  | xxx |
| Machinery | xxx |  |
|  | xxxx | xxxx |

| Dr side | Cr side |
|---|---|
| A, E | C, L, I |

Classwork 2

## Purchase

|  | $ |  | $ |
|---|---|---|---|
| Cash | 100 |  |  |
| Cash | 20 |  |  |

## Debtor

| | $ | | $ |
|---|---|---|---|
| | 700 | | |

## Sales

| | $ | | $ |
|---|---|---|---|
| | | Cash | 250 |
| | | Cash | 500 |
| | | Debtor | 700 |

## Cash

| | $ | | $ |
|---|---|---|---|
| Bal b/d | 100 | | |

Balance the above ledger account and then prepare trial balance.

## Trial Balance

|  | Dr. | Cr. |
|---|---|---|
|  | $ | $ |
| Purchase |  |  |
| Sale |  |  |
| Cash |  |  |
| Debtor |  |  |

## [04] The use of Trial Balance

There are several uses to which a trial balance may be put. These are:

◎To check that the books 'balance'.

◎To ascertain the net amount of the error(s), if any have been made, as represented by the difference between the two sides.

## [05] Types of accounts

(i)     Personal account — Debtors, Creditors, Capital

(ii)    Nominal account — Incomes, Expenses

(iii)   Real account - others

## [06] Examination

| 2003 |  | S3 |
|---|---|---|
| 2004 |  |  |
| 2005 | S2 |  |

# Chapter 4

## Final Accounts

## Content

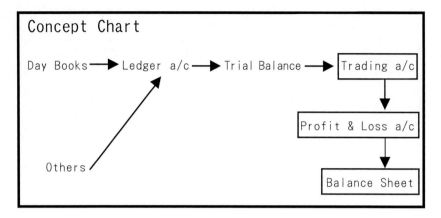

## Concept Chart

Day Books ➤ Ledger a/c ➤ Trial Balance ➤ Trading a/c

Trading a/c → Profit & Loss a/c → Balance Sheet

Others ↗

## [01]  Gross Profit and Net Profit

Trading Account     +     Profit & Loss Account

Gross Profit                Net Profit

Gross Profit = Net Sales − Cost of Goods Sold

Net Sales = Total Sales − Return Inwards

## [02]  Equation of "Cost of Goods Sold"

Cost of Goods Sold
= Opening Stock + (Purchase − Return Outward + Carriage Inwards)
  − Closing Stock

## [03]  Items of Trading Account

Items of " T" → Related to  'Stock'!

## [04] Vertical style for trading accounts

| Trading Account for the year ended 31 December 20X7 | | |
|---|---|---|
| | $ | $ |
| Sales | | 3,850 |
| Less : Cost of goods sold: | | |
| Opening Stock | 100 | |
| Purchases | 2,900 | |
| Closing stock | (400) | 2,600 |
| Gross profit | | 1,250 |

## Classwork 1

| | |
|---|---|
| Sale | 200 |
| Opening Stock | 1500 |
| Purchase | 700 |
| Closing Stock | 1000 |

## [05] Vertical style for profit and loss accounts

NP = GP + other income - expenses

| Profit and Loss A/C for the year ended 31 December 20X7 | | |
|---|---|---|
| | $ | $ |
| Gross profit | | 1,250 |
| Less: Expenses: | | |
| Rent | 240 | |
| Lighting | 150 | |
| Heating | | |
| Net profit | | 800 |

## Classwork 2

Rent    200
Rate    100

## [06]   Vertical style for trading and profit and loss A/C

| Trading and Profit and Loss A/C for the year ended 31 Dec 20X7 | | |
|---|---|---|
| | $ | $ |
| Sales | | 3,850 |
| Less: Cost of goods sold: | | |
| Purchases | 2,900 | |
| Closing stock | (300) | 2,600 |
| Gross profit | | 1,250 |
| Less: Expenses: | | |
| Rent | 240 | |
| Lighting | 150 | |
| Heating | 60 | 450 |
| Net profit | | 800 |

## [07]   Classification

| Item | Class | Definition | Examples |
|---|---|---|---|
| Assets | Fixed Assets | · Long life use<br>· Not for resale | Land, buildings |
| | Current Assets | · Short life use<br>· For resale | Stock, debtors |
| Liabilities | Long Term Liabilities | · More than one yr | Bank loan |
| | Current Liabilities | · Less than one yr | Creditors |

## [08]   Vertical Style balance sheets

$$A = C + L$$

A                                     C + L

### Balance Sheet as at 31 December 20X7

|  | $ | $ | $ |
|---|---|---|---|
| Fixed Assets | Cost | Acc. Depr | NBV |
| Fixtures and fittings | 500 | 0 | 500 |
|  | 500 | 0 | 500 |
|  |  |  |  |
| Current Assets |  |  |  |
| Stock | 300 |  |  |
| Debtors | 680 |  |  |
| Bank | 1,510 |  |  |
| Cash | 20 | 2,510 |  |
|  |  |  |  |
| Less Amount due within one year |  |  |  |
| Creditors |  | 910 |  |
| Net Current Assets (CA − CL) |  |  | 1,600 |
|  |  | * | 2,100 |
|  |  |  |  |
| Financed by: |  |  |  |
| Capital |  |  |  |
| Cash introduced |  |  | 2,000 |
| Net profit for the year |  |  | 800 |
|  |  |  | 2,800 |
| Drawings |  |  | (700) |
| Balance as at 31 Dec 20x7 |  |  | 2,100 |

* $2100 is the value of company in this moment.

42

## [09]   Equation

Final account = Trading & Profit & Loss A/C + Balance Sheet

## [10]   Function of Financial Statements

| Financial Statements | Purpose |
|---|---|
| Trading, Profit & Loss a/c | Calculate the net profit |
| Balance Sheets | To show the assets and liabilities |

## [11]   Examination

| 2003 | S2 |    | S4 |
|---|---|---|---|
| 2004 | S2 | S3 | S4 |
| 2005 | S2 | S3 | S4 |
| 2006 | S2 |    |    |

## [12]   Exercise

Please correct the format and figure

|  | $ | $ |
|---|---|---|
| Sales | 10,000 |  |
| Less: Cost of Good Sold |  |  |
| Opening Stock |  | 1,000 |
| Closing Stock | -500 |  |
| Purchase |  | 200 |
| Gross Profit | 9,000 |  |

# Chapter 5

## Bad Debts and Provision for Bad Debts

## Content

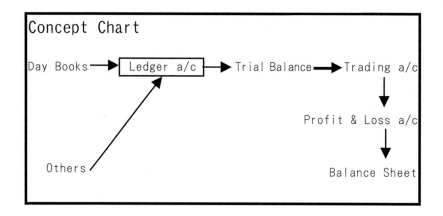

## Concept Chart

Day Books ➔ Ledger a/c ➔ Trial Balance ➔ Trading a/c

↓

Profit & Loss a/c

Others ↗

↓

Balance Sheet

## [01]   Introduction

### Balance Sheet (extract)

| Current Assets | $ | $ | $ |
|---|---|---|---|
| Stock | X | | |
| Debtors | X | | |

How to count it?

## [02]   Bad Debt

When the firm finds that the debtor would not paid, that debt should be written off.

The accounting entries are:

| Accounting entries | Explanation |
|---|---|
| Debit: Bad Debts account | To transfer the amount of uncollected debt to Bad Debts account |
| Credit: Debtors account | To reduce the amount of the debtor who is unable to paid |

45

## [03]　Why a business might create a provision for doubt debts

◎Prudence Concept – can avoid overstating profits and debtor

◎Matching Concept – can relate any loss for the period with the related sales in the same period

### Classwork 1

We sold $1,000 goods to Mark on 1 January 20X7, but he became bankrupt. On 1 February 20X7 we sold $2,000 goods to Jimmy.　He managed to pay $200 on 17 May 20X7, but it became obvious that he would never be able to pay the final.

| Mark | |
|---|---|
| | |
| | |

| Jimmy | |
|---|---|
| | |
| | |

| Bad Debt | |
|---|---|
| | |
| | |

## [04]　The posting for provision for bad debts

Provision for bad debts shows the estimated amounts for debtors, who will be unable to pay.

### Provision for Bad Debts

| | $ | | | $ |
|---|---|---|---|---|
| 2001 Bal c/d | A | 2001 | Profit & loss a/c | X |
| | | | | |
| 2002 Profit & loss a/c | X | 2002 | Bal b/d | A |
| | | | | |
| Bal c/d | B | | | |
| | X | | | X |

### Balance Sheets (Extracts)

| | $ | $ |
|---|---|---|
| 2001 Debtors | X | |
| Less Provision for bad debts | A | X |
| 2002 Debtors | X | |
| Less Provision for bad debts | B | X |

## [05]　Increasing the provision　(Bal c/d > Bal b/d)

The double entry will be:
　　　　Debit : Profit and Loss Account
　　　　Credit : Provision for Bad Debts account
【Expenses】

## [06]  Reducing the provision (Bal c/d < Bal b/d)

The double entry is:

        Debit : Provision for Bad Debts account
        Credit : Profit and Loss Account

【Income】

## Classwork 2

| Year | Credit Sales | Bad Debt | Net Debtor | % |
|---|---|---|---|---|
| 1 | 10,000 | 100 | | 5% |
| 2 | 15,000 | 100 | | 5% |
| 3 | 20,000 | 100 | | 1% |

| Debtor | |
|---|---|
| | |

| Bad Debt | |
|---|---|
| | |

| Profit & Loss account (extract) | | | |
|---|---|---|---|
| | $ | $ | $ |
| | | | |

| Balance Sheet (extract) | | | |
|---|---|---|---|
| | $ | $ | $ |
| | | | |

## [07]  Different between bad debt and provision for bad debt

|  | Bad Debt | Provision for bad debt |
|---|---|---|
| Types | Expenses | Expenses / Incomes |
| Placing | PL / CA | PL / CA |
| Accuracy | Known with certainly | Estimate of possible losses |

## [08]  Bad debts recovered

(a) Period: more than one year

First, reinstate the debt by making the following entries:

Debit: Bank account

Credit: Debtor


Debit: Debtor

Credit: Bad Debt Recover (incomes)

With the amount received.


(b) Period: on the current financial year

Dr: Bank

Cr: Bad Debt

## Classwork 3

On 5 March 20x7, $2,500 was received from Mr So in settlement of the debt which had previously been written off as bad in 20x6.

|  |  |
|---|---|
|  |  |
|  |  |

## [09]  Provision for Discounts

If it is desired to show the debtors' balance at a more realistic figure, another provision account namely "Provision for Discount Allowed" can be made.

### Classwork 4

Debtor $100,000, provision for doubtful debts $2,000, discount for provision is 5%, bad debt $10,000

| Balance Sheet (extract) | | | |
|---|---|---|---|
| | $ | $ | $ |
| Debtors | | | |
| Less :   Provision for doubtful debts | | | |
| Provision for discounts on debtor | ___ | ___ | |

The firms find that the creditors' balance does not represent the amount which will paid.  A Provision for Discount Received is made.

Dr: Provision for discount received

Cr: Profit & Loss a/c

## [10]  Examinations

| 2003 | S3 | |
|---|---|---|
| 2004 | S3 | S4 |
| 2005 | | S4 |

# Chapter 6

Depreciation & Disposal

## Content

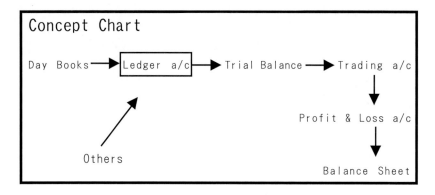

## Concept Chart

Day Books ➡ Ledger a/c ➡ Trial Balance ➡ Trading a/c

⬇ Profit & Loss a/c

Others ⬆

Balance Sheet

## [01] Introduction

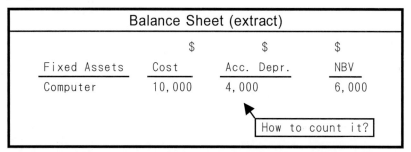

### Balance Sheet (extract)

| Fixed Assets | Cost | Acc. Depr. | NBV |
|---|---|---|---|
| | $ | $ | $ |
| Computer | 10,000 | 4,000 | 6,000 |

How to count it?

## [02] Depreciation of Fixed Assets

· Cost of the fixed asset consumed
· It is an expense for services consumed
· Matching concept

## [03] Causes of Depreciation

a. Physical depreciation
  Wear & tear          Rust, rot and decay
b. Economics Factors
  Obsolescence         Inadequacy
c. The time factor
d. Depletion
The wasting away of an asset as it is used as it is used up.

## [04]  Land & Building

a. Land does not require a provision for depreciation (Except mines and quaries)

b. Buildings require a provision for depreciation

## [05]  Methods of calculating depreciation charges

a. Straight Line Method — at an equal amount each year.

## Classwork 1

If we bought a motor van for $20,000, thinking that we would keep if for 5 years and then sell it for $5,000.

Depreciation
= (Cost Price-Disposal Value)/No of estimated year of use

=

=

b. Reducing balance method — based on the net book value (NBV) of the asset brought forward from the previous year.

(Net Book Value=Cost — Accumulated Depreciation)

## Classwork 2

If we bought a motor van for $20,000, thinking that we would keep if for 5 years and then will sell it for $5,000, depreciation rate is 20%.

| Cost | 20,000 |
|---|---|
| First Year Depreciation ($20,000 x          ) | |
| Net book value at the end of the first year | _____ |
| Second Year Depreciation ($         x         ) | |
| Net book value at the end of the second year | |
| Third Year Depreciation ($       x        ) | _____ |
| Net book value at the end of the Third year | |
| | ═══════ |

## [06]  Double entry of Depreciation

The double entry is:

Dr: Profit & Loss a/c

Cr: Provision for depreciation a/c

## Classwork 3

A machine is bought for $10,000 on 1 Jan 20X5.  It is to be depreciated at the rate of 10% using the reducing balance method, prepare 2 years' account.

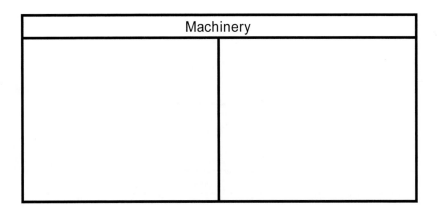

| Machinery | |
|---|---|
| | |

| Provision for Depreciation | |
|---|---|
| | |

## [07]  Disposal

Upon the sale of a fixed asset, we will want to remove it from our ledger accounts.

### Classwork 4

Use the above data, sell the machinery $3,500 at 2 Jan 20X7.

| Disposal | |
|---|---|
| | |
| | |
| | |

### Classwork 5

Sell the machinery $9,000 at 2 Jan 20X7.

| Disposal | |
|---|---|
| | |
| | |
| | |

| | Dr | Cr |
|---|---|---|
| Profit on Disposal | | |
| Loss on Disposal | | |

## [08]  Trade In (Additional Information )

| 賣資產五部曲: 成、舊、錢、換、走 | |
|---|---|
| 1. Cost of FA 成 本 | Dr: Disposal        Cr: FA (Cost) (↓) |
| 2. Accumulated Depreciation 累 積 折 舊 | Dr: Provision for Depreciation(↓)    Cr: Disposal |
| 3. Money Received 收 錢 | Dr: Cash / Bank(↑)    Cr: Disposal |
| 4. Trade in value 以 機 換 機 | Dr: FA(↑) (trade in value)    Cr: Disposal (same nature with cash receipt) |
| 5. Profit or Loss 計 完 賺 蝕 就 走 | Profit<br>Dr: Disposal      Cr: Profit & Loss<br>Loss<br>Dr: Profit &      Cr: Disposal |

## [09]  Examinations

| 2003 | S2 | S3 | |
|---|---|---|---|
| 2004 | S2 | S3 | S4 |
| 2005 | S2 | S3 | |

# Chapter 7

## Cash Book

## Content

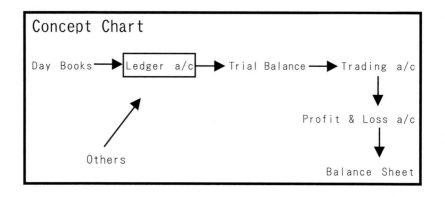

## Concept Chart

Day Books ➔ Ledger a/c ➔ Trial Balance ➔ Trading a/c

Others ➔

Profit & Loss a/c

Balance Sheet

## [01]  Introduction

Cash Book  =  Bank a/c + Cash a/c + Discount Received + Discount Allowed

Dr → ↑                    Cr → ↓

## [02]  Format

### Cash Book

| Year | | F | Discount | Cash | Bank | Year | F | Discount | Cash | Bank |
|------|------|---|----------|------|------|------|---|----------|------|------|
| Date | Items | | $ | $ | $ | Date | | $ | $ | $ |

Total of this column
is transferred to Dis-
counts Allowed a/c.

Total of this column is
transferred to Discounts
Received a/c.

◎The bank column contains details of the payment and receipts made by cheques

◎The cash column contains details of the payment and receipts made by cash

◎When the firm received the bank statement; check against the bank column in cash book

## [03]  The use of folio columns

Folio columns used for entering reference numbers.

### Classwork 1

An entry of receipt of a cheque from May whose account was on page 5 of the sales ledger, and the cheque recorded on page 7 of the cash book, would use the folio column as follows:

◎In the Cash Book : in the folio column would appear SL 5

◎In the Sales Ledger : in the folio column would appear CB 7

| Year | | F | Dis | Cash | Bank | Year | F | Dis | Cash | Bank |
|------|------|---|-----|------|------|------|---|-----|------|------|
| Date | Items | | $ | $ | $ | Date | | $ | $ | $ |
| Nov25 | Sales | | | | | | | | | |

### Sales

| 20X7 | | F | $ | 20X7 | | F | $ |
|------|---|---|---|------|---|---|---|
| | | | | Nov 25 | | | 800 |

The advantages of folio columns

◎Speeds up reference to the other books

◎Make easier to find errors when a transaction has only been entered once.

## [04]  Cash Discount

Cash Discount is an allowance given to customers who pay early.

## [05]  Discounts Allowed & Discounts Received

It is usually expressed as a percentage, e.g 7%

| | |
|---|---|
| Discounts Allowed | to customers when they pay early (expenses) |
| Discounts Received | from suppliers when the firm pay quickly. (revenue) |

Alvin owned us $1,000.  He paid us $950 and we allowed him $50 for the reward of quickly pay. We paid $980 for settle the debt of $1,000.

### Cash Book

| | F | Discount | Cash | Bank | | F | Discount | Cash | Bank |
|---|---|---|---|---|---|---|---|---|---|
| | – | $ | $ | $ | Creditor | | $ | $ | $ |
| Debtors | | 50 | 950 | | | | 20 | 980 | |

| | |
|---|---|
| Transfer to the debit of discount allowed a/c | Transfer to the credit of discount received a/c |

The discount columns are NOT part of the double entry.

## [06]  Cash paid into bank (cash decrease, bank increase)

| 20x7 | | F | Discount | Cash | Bank | 20x7 | | F | Discount | Cash | Bank |
|---|---|---|---|---|---|---|---|---|---|---|---|
| | | | $ | $ | $ | | | | $ | $ | $ |
| Nov25 | Cash | C | | | 800 | Nov25 | Bank | | | 800 | |

Contra is where both the debit and credit entries are shown in the cash book.

Contra sign "C" should be inserted in the "F" column

## [07]  Balancing the Cash Book

To balance off an account:

Cash & Bank Column

1. same as the other ledger accounts

2. find out balance c/d

Discount Column

1. add up the amount

Classwork 2

### Cash Book

| | | Discount | Cash | Bank | | | | Discount | |
|---|---|---|---|---|---|---|---|---|---|
| | | Cash | Bank | | | | | | |
| | | $ | $ | $ | | | | $ | $ | $ |
| Jan 1 | Sales | | 100 | | 100 | Jan 1 | Creditors | | 30 | 200 |
| | Debtors | — | —20— | | 980 | | Bal c/d | — | — | —45 |
| 255 | | | | | | | | | | |

## [08]  Comparison of Trade Discount and Cash Discount

30

| | Trade Discount | Cash Discount |
|---|---|---|
| Nature | A deduction from list price | A deduction of the sum to be paid |
| Purpose | To a trader buying goods in large quantities | To encourage customers to pay promptly |
| Double Entry | Only post in day books | Double entry system |

## [09]   Bank Overdrafts (Expenses > Income + Bal b/d)

The balance becomes a credit one. It is current liabilities (CL).

## Classwork 3

Amy draw $2000 form her bank a/c of Hong Seng Bank by ATM.

### Cash Book

| 20X7 | F | D | C | B | 20X7 | F | D | C | B |
|------|---|---|---|---|------|---|---|---|---|
| May1 | | | | | May1 | | | | |

## [10]   Examination

| 2004 | S2 | |
|------|----|----|
| 2005 | S2 | S3 |

# Chapter 8

## Petty Cash Book

## Content

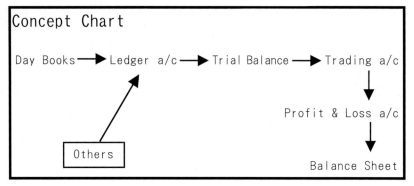

```
Concept Chart

Day Books ──▶ Ledger a/c ──▶ Trial Balance ──▶ Trading a/c
                     ▲                                │
                     │                                ▼
                     │                        Profit & Loss a/c
              ┌──────────┐                            │
              │  Others  │                            ▼
              └──────────┘                     Balance Sheet
```

## [01]  Division of the cash book

Petty cash book is a cash book for posting small cash payments.

## [02]  Why we need it?

◎ Many small cash payment to be made

◎ Handling the small payments given to a junior book-keeper.

## [03]  Voucher

Voucher is a form that shows what a payment. It included the following items:

◎ Reference Number

◎ Date

◎ For what required

◎ Amount

◎ Signature

◎ Passed by

### Petty Cash Book

| Receipts | Date | Detail | Voucher No. | Total | Postage | Clean-ing | Travel-ing | Sundries | Ledger Accounts |
|----------|------|--------|-------------|-------|---------|-----------|------------|----------|-----------------|
| $ |  |  |  | $ | $ | $ | $ | $ | $ |

## [04] Imprest system

It is a system, which a refund is made of the total paid out in a period.

### Classwork 1

| 20X0 | | |
|---|---|---|
| | The cashier gave the petty cashier | 200 |
| | The petty cashier paid out in 20x0 | 60 |
| | Petty cash now in hand | 140 |
| | The cashier now gives the petty cashier | |
| | the amount spent | 60 |
| | Petty cash in hand at the end of 20x0 | 200 |
| | | |
| 20X1 | | |
| | The petty cashier paid out in  20x1 | 56 |
| | Petty cash now in hand | 144 |
| | The cashier now gives the petty cashier | |
| | the amount spent | |
| | Petty cash in hand at the end of  20x1 | |

### Classwork 2

Cathy Wong keeps her petty cash on the imprest system – amount being $1,000. The petty cash transactions for the month of March 20X7 were as follows:

| Mar | |
|---|---|
| 1 | Petty cash in hand $800. |
| 1 | Petty cash restored to imprest amount |
| 6 | Bought pencil $60. |
| 7 | Paid wages $90. |
| 20 | Paid to Mr Lee, a creditor, $100. |
| 21 | Paid wages $70. |
| 25 | Bought envelopes $15. |
| 28 | Bought notepaper $20. |

# Cathy Wong

## Petty Cash Book

| Receipts | Date | Details | Total | Stationery | Wages | Postage | Ledger |
|---|---|---|---|---|---|---|---|
| $ | 20X7 | | $ | $ | $ | $ | $ |
| 800 | Mar 1 | Balance b/f | | | | | |
| 600 | 1 | Cash | | | | | |
| | 6 | Stationery | 60 | 60 | | | |
| | 7 | Wages | 90 | | 90 | | |
| | 20 | Mr. Lee | 100 | | | | 100 |
| | 21 | Wages | 70 | | 70 | | |
| | 25 | Envelopes | 15 | | | 15 | |
| | 28 | Notepaper | 20 | 20 | | | |
| | | | | 80 | 160 | | 100 |
| | 31 | Balance c/d | | | | | |
| 1,000 | | | 1,000 | | | | |
| | Apr 1 | Balance b/d | | | | | |
| | 1 | Cash | | | | | |

## [05]　Imprest System in different period

### 1.Information

```
Mar  1   Received a cash float $600
Mar  2   Petrol $70
Mar  3   CD $60
Mar  4   Postage $90
Mar  5   A4 paper $60
```

## 2.Reimbursement at the boxed(end) of a period

### Petty Cash Book

| Receipts | F | Date | Details | Voucher | Total | Motor Vehicles | Postage | Stationery |
|---|---|---|---|---|---|---|---|---|
| $ 600 | | 1/3 | Cash | | $ | $ | $ | $ |
| | | 2 | Petro | I | 70 | 70 | | |
| | | 3 | CD | 2 | 60 | | | 60 |
| | | 4 | Postage | 3 | 90 | | 90 | |
| | | 5 | Paper | 4 | 60 | | | 60 |
| | | | | | 280 | 70 | 90 | 120 |
| 280 | | 31 | Cash | | | | | |
| | | | Bal c/d | | 600 | | | |
| 880 | | | | | 880 | | | |

## 3.Reimbursement at the boxed(beginning) of a period

### Petty Cash Book

| Receipts | F | Date | Details | Voucher | Total | Motor Vehicles | Postage | Stationery |
|---|---|---|---|---|---|---|---|---|
| $ 600 | | 1/3 | Cash | | $ | $ | $ | $ |
| | | 2 | Petro | I | 70 | 70 | | |
| | | 3 | CD | 2 | 60 | | | 60 |
| | | 4 | Postage | 3 | 90 | | 90 | |
| | | 5 | Paper | 4 | 60 | | | 60 |
| | | | | | 280 | 70 | 90 | 120 |
| 600 | | 31 | Bal c/d | | 320 | | | |
| 320 | | | | | 600 | | | |
| 280 | | | | | | | | |

## [06]    Examinations

| 2003 | | S4 |
|------|------|------|
| 2004 | | |
| 2005 | | S4 |
| 2006 | S2 | |

**Petty Cash Book**

# Chapter 9

## Organisation

## Content

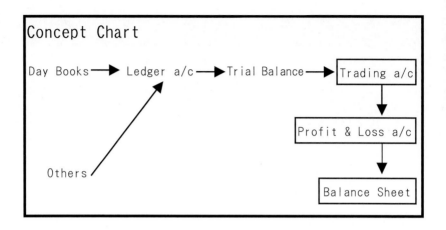

```
Concept Chart

Day Books ──▶ Ledger a/c ──▶ Trial Balance ──▶ | Trading a/c |
                        ▲                                  │
                       ╱                                   ▼
                      ╱                         | Profit & Loss a/c |
                     ╱                                     │
          Others ──╱                                       ▼
                                                | Balance Sheet |
```

## [01]  Introduction

Club and society provide services for members, not for profit-making.

## [02]  Receipts and Payments Account

### Receipts and Payments Account for the year ended

### 31 December 20X6

|  | $ |  | $ |
|---|---|---|---|
| Balance b/d | 236 | Groundsman's wages | 728 |
| Subscriptions received | 1,148 | Sports stadium expenses | 296 |
| Rent received | 116 | Committee expenses | 58 |
|  |  | Printing and stationery | 33 |
|  |  | Balance c/d | 310 |
|  | 1,500 |  | 1,500 |

| Profit Making Firm | Non-profit Making Organization |
|---|---|
| 1. Cash Book | 1. Receipts and Payments Account |

## [03] Income and Expenditure Accounts

Income and expenditure account is an account for a non-profit making organisation to find the surplus or deficit made during a period.

| Profit Making Firm | Non-profit Making Organization |
|---|---|
| 1. Trading account | 1. Trading a/c |
| 2. profit and loss account | 2. Income & Expenditure a/c |
| 3. Net profit | 3. Surplus |
| 4. Net loss | 4. Deficit |

## [04] Accumulated Fund

Accumulated Fund = Assets - Liabilities = Capital

| Profit Making Firm | Non-profit Making Organization |
|---|---|
| 1. Cash Book | 1. Receipts and Payments Account |
| 2. Drawing | 2. Not available |

## Classwork 1

Sport Equipment $8,000, Club Premises $22,000, Furniture & Fittings $5,000, Rates & Insurances $200, Bank $2,000, Creditor $1,000, $ Bar Stock $2,000

Use the above information to prepare the statement of affairs.

| Statement of Affairs | | |
|---|---|---|
| | $ | $ |
| | | |
| | | |
| | | |
| | | |

## [05]　Drawing up Final Accounts

A separate trading account is to be prepared for a bar, where is made for one kind of income.

## Classwork 2

| Trial Balance as at 31 December 20X7 | | |
|---|---|---|
| | $ | $ |
| Equipment | 8,000 | |
| Premises | 29,000 | |
| Subscriptions | | 6,000 |
| Wages | 4,000 | |
| Furniture and fittings | 5,000 | |
| Rates and Rent | 1,900 | |
| Sundry expenses | 600 | |
| Accumulated fund 1 January 20X7 | | [      ] |
| Donations | | 300 |
| Telephone ,postage & Printing | 500 | |
| Bank | 2,000 | |
| Bar purchases | 9,500 | |
| Creditors | | 1,000 |
| Bar sales | | 14,000 |
| Bar stocks 1 January 20X7 | 2,000 | |

## Classwork 3

The following information is also available: (i) Bar stocks at 31 December 20X7 $2,362. (ii) Provision for depreciation: sports equipment $1,700; Furniture and fittings $1,315.

### Bar Trading Account for the year ended 31 December 20X7

| | $ | $ |
|---|---|---|
| Sales | | |
| Less Cost of goods sold: | | |
|     Opening stock | | |
|     Purchases | | |
|     Less Closing stock | ——— | ——— |
| Bar profit to income and expenditure account | | ═══ |

### Income and expenditure Account for the year ended 31 December 20X7

| Income | $ | $ | $ |
|---|---|---|---|
| Gross profit from bar | | | ◄ |
| Subscriptions | | | |
| Donations | | ——— | |
| | | | |
| Less Expenditure | | | |
| Wages | | | |
| Rates and insurance | | | |
| Telephone and postage | | | |
| General expenses | | | |
| Provision for Depreciation: | | | |
| · Furniture and fittings | | | |
| · Sports equipment | ——— | ——— | ——— |
| Surplus of income over expenditure | | | ═══ |

## Balance Sheet as at 31 December 20x7

|                          | $ Cost | $ Acc. Depr. | $ NBV |
|--------------------------|--------|--------------|-------|
| **Fixed assets**         |        |              |       |
| Club premises            |        |              |       |
| Furniture and fittings   |        |              |       |
| Sports equipment         |        |              |       |

**Current Assets**
Bar stocks
Cash in bank

**Current Liabilities**
Creditors for bar supplies
Net current assets

Financed by:
**Accumulated fund**
Balance at 1 January 20x7
Add Surplus
Balance as at 31 December 20x7

Surplus =      >
Deficit =      <

75

# [06] Subscriptions

## Subscriptions

| 20X7 | | | $ | 20X7 | | | $ |
|------|---|---|---|------|---|---|---|
| Jan 1 | In arrear b/d | (1) | 360 | Jan 1 | In advance b/d | (2) | 80 |
| Dec 31 | I & E (Bal fig) | | *7,220 | Dec 31 | Bank | (4) | 7,420 |
| | In advance c/d | (3) | 140 | | In arrear c/d | (5) | 220 |
| | | | 7,720 | | | | |
| 20X8 | | | | 20X8 | | | |
| Jan 1 | In arrear b/d | (5) | 220 | Jan 1 | In advance b/d | (3) | 140 |

# [07] Donations

Any donations received are shown as income in the year that they are received.

# [08] Entrance Fees

New members often need to pay an entrance fee when they join. Entrance fees are normally included as income in the year that they are received. Sometimes it would treat as accumulated fund at examination.

# [09] The double entry of donations and entrance fees

| Description | Double Entry |
|-------------|--------------|
| Donation received or entrance fee received | Dr: Cash<br>Cr: Donation / Entrance Fee Received |

## [10]  Life Membership

Sometimes members can pay one amount for membership, and then this membership will last for their lifetime.

## [11]  Different between I & E and R & P

| | Receipts & payment a/c | Income & Expenditure a/c |
|---|---|---|
| **Nature** | Record the cash in and out in the period briefly | Calculate the surplus and deficit |

## [12]  Examinations

| 2003 | | | |
|---|---|---|---|
| 2004 | S2 | | |
| 2005 | | | |

# Chapter 10

## Bank Reconciliation Statement

## Content

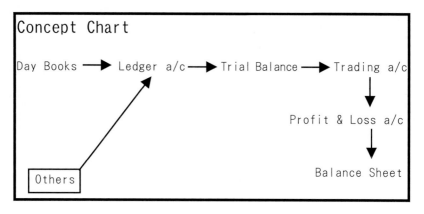

## [01] The purpose of the bank reconciliation statement

◎ explain the reasons for the differences
◎ to identify errors and omissions in bank account and bank statement.

## [02] Reasons for differences

Due to the timing difference, omissions and errors made by the bank or the firm itself. The balances of the bank statement by bank and the bank account by firm did not agree.

| Sources | Items |
|---|---|
| Internal | Uncredited Cheque |
| | Error on Cash Book |
| External | Credit Transfer |
| | Standing Order |
| | Bank Interest |
| | Bank Charge |
| | Bank Lodgment |
| | Error on Bank Statement |
| Third Parties | Unpresented Cheque |
| | Dishonoured Cheque |

## (i)   Dishonoured cheques

They are cheques deposited but subsequently returned by the bank due to the failure of the drawer to pay.

### Example

Jan 1   A debtor, Peter give us a cheque $100

Jan 2   Bank tell us the Peter's cheque dishonourable

| Cash Book | | | |
|---|---|---|---|
| | $ | [2] | $ |
| Jan 1 Debtor — Peter [1] | | Jan 2 Debtor — Peter | |
| | 100 | (Dishonourable) | 100 |

### Classwork 1

Feb 1   Peter give us $1,000 cheque and $200 cash

Feb 5   It has dishonorable for Peter

| Cash Book | |
|---|---|
| | |

## (ii) Bank charges

They are charges made by the bank to the company for banking services provided.

## (iii) Standing orders S/O

The instructions from the firm to the bank to make regular payments such as rent.

## (iv) Direct debits D/D

The instructions from the firm to the bank to make non-regular payments such as electricity.

## (v) Credit transfers C/T

They are collections from customers directly through the bank.

## (vi) Interest allowed by the bank

They are interest received for fixed or time deposits.

## (vii) Uncredited items

The deposits occurred too close to the cut-off date of the bank statement and so do not appear on the statement such as 31 Dec. They will appear on the next statement.

## (viii) Unpresented cheques

They are cheques issued by the firm that have not yet been presented. The firm has not paid any.

## [03]    Nature of the cash book and bank statement

Cash Book = Asset

### Cash Book

| + | $ | − | $ |
|---|---|---|---|
| A debit represents an increase | | A credit represents a decrease | |

The balance as per the bank statement is a liability to the bank, so:

| | Bank statement | | |
|---|---|---|---|
| | Dr | Cr | Balance |
| | $ | $ | $ |
| | Withdrawal | Deposit | Represents the amount owed to the film |
| | — | + | |

## [04]    Prepare the bank reconciliation statement

| First Step (Updated Cash Book) | | | |
|---|---|---|---|
| | $ | | $ |
| Bal b/d | X | Dishonoured cheques | X |
| Credit transfers | X | Bank charges | X |
| Bank interest received | X | Standing orders | X |
| Wrong credit by the bank | X | Direct Debit | X |
| | | Wrong debit by the bank | X |
| | | Balance c/d | A |
| | X | | X |

| Second Step (Bank reconciliation statement as at XXX) | |
|---|---|
| | $ |
| Balance as per cash book | A |
| Add: Unpresented cheques | X |
| | X |
| | |
| Less : Uncredited items | X |
| | X |
| Balance as per bank statement | X |

# Classwork 2

| Cash Book | | | | | |
|---|---|---|---|---|---|
| 20x7 | | $ | 20x7 | | $ |
| Jun1 | Balance b/f | 2,000 | Jun5 | D. Wong | 150 |
| 7 | B. Moon | 160 | 12 | J. Yau #1004 | 400 |
| 11 | Mr Ng 200 | 200 | 17 | Peter #1005 | 100 |

| Bank Statement | | | | |
|---|---|---|---|---|
| 20x7 | | Dr.$ | Cr.$ | Balance $ |
| Jun 1 | Balance b/f | | | 2000 |
| 7 | Cheque | | 160 | 2,160 |
| 8 | Wong | 150 | | 2,010 |
| 9 | Bank Charge | 100 | | 1,910 |
| 10 | Mr So | | 1,000 | 2,910 |

You are required to:

(a) Write the cash book up-to-date to take the above into account.

(b) Draw up a bank reconciliation statement as on 30 June 20x7 as begin as update cash book balance.

(c) Draw up a bank reconciliation statement as on 30 June 20x7 as begin as bank statement balance.

## [05]   Bank Overdrafts

The process is the same as those needed when the account is not overdrawn.

# Classwork 3

## Cash Book

| | | | |
|---|---|---|---|
| Bal b/d | 100 | Rent S/O | 1000 |
| May C/T | 50 | Electricity D/D | 100 |
| Bal c/d | | Bank Charge | 20 |

## Bank Reconciliation Statement

| | $ | $ |
|---|---|---|
| Balance as per Cash Book | | |

## [06]　Dishonoured Cheque

After banked the cheque, the customer's bank will not pay the amount due on the cheque, the customer's bank has failed to "honour' the cheque.

# Classwork 4

In 1 Jan 20x7, received $1,000 from May.

In 3 Jan 20x7, our bank tell us May's account has not sufficient funds.

## Bank

# Debtor-May

## [07] Examination

| 2003 | S2 |    |
|------|-----|-----|
| 2004 |     | S3 |
| 2005 | S2  | S3 |
| 2006 | S2  |    |

# Chapter 11

Journal

## Content

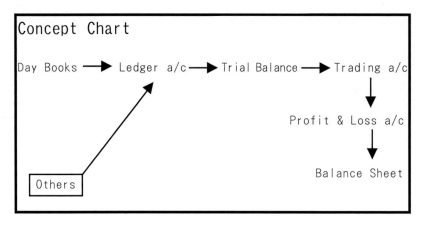

Concept Chart

Day Books ➤ Ledger a/c ➤ Trial Balance ➤ Trading a/c

Profit & Loss a/c

Balance Sheet

Others

## [01] The Nature of Journal

Journal is used to record those transactions, that cannot go the subsidiary day books, i.e. Cash Book, Sales Journal, Purchase Journal, Return Inwards Journal, Return Outwards Journal, and Petty Cash Book. It is NOT part of the Double Entry System.

| Journal | | | | Page |
|---------|---|---|---|---|
| Date | Particulars | F | Dr | Cr |
| | | | $ | $ |
| | | | | |

## [02] Transaction to be recorded in Journal

◎ Opening entries

◎ Purchase and sale of fixed assets on credit

◎ The correction of errors

◎ All other transfer that cannot record in any day books

## Classwork 1

Mr So invested Premises $350,000, Motor Van $60,000 and Cash $28,000 to start his business on January 1, 20x7. To begin writing the books, these items must be entered into the Journal.

| Journal | | | P.1 |
|---|---|---|---|
| Date | Particulars | F | Dr Cr |
| 20x7 | | | $ $ |
| Jan 1 | | | |
| | | | |
| | | | |
| | Being Assets and Capital at | | |
| | opening new set of books | | |

## Classwork 2

Maria Company after being in business for a number of years, decided to keep a proper set of double entry books. On January 1, 20x7, its assets and liabilities were as follows:

◎ Assets :

Fixtures and Fittings $20,000 , Office Equipment $10,000 , Motor Vehicle $26,000 , Debtors $17,000 , Cash $10,000

◎ Liabilities:

Creditors $21,000

| Journal | | | P.1 | |
|---------|---|---|-----|---|
| Date | Particulars | F | Dr | Cr |
| 20x7 | | | $ | $ |
| Jan 1 | | | | |
| | Being Assets, Liabilities and Capital at opening new set of books | | | |

## Classwork 3

A typewriter was bought on credit from Kowloon Co. for $4,000 on January 3, 20x7.

| Journal | | | P.1 | |
|---------|---|---|-----|---|
| Date | Particulars | F | Dr | Cr |
| 20x7 | | | $ | $ |
| Jan 3 | | | | |
| | Purchase of typewriter on credit. | | | |

## Classwork 4

On April 16, 20x7, an office desk, bought for use in the firm for $1,500, was sold on credit to Kennedy Co. for $1,300 and no provision for depreciation had been made.

| Journal | | | | P.3 |
|---|---|---|---|---|
| Date | Particulars | F | Dr | Cr |
| 20x7 | | | $ | $ |
| Apr 16 | | | | |
| | Sale of office desk on credit | | | |
| | Sold office equipment to Kennedy Co. on credit | | | |
| | Loss on sales of office equipment | | | |

## Classwork 5

Stock $500 was taken out as sample for promotion on May 16, 20x7, $100 for promotion and the balance for private use.

| Journal | | | | P.7 |
|---|---|---|---|---|
| Date | Particulars | F | Dr | Cr |
| | | | $ | $ |
| 20x7 | | | | |
| | Stock withdrew as sample for promotion | | | |

## [03]  Errors Affecting Trial Balance Agreement

◎ forgetting to post either the sides.
◎ wrong calculation in the day book.
◎ wrong calculation the balance of an account
◎ Debit and credit amounts are different
◎ Post an entry into the wrong side of an account.

## [04]  The Effect of Errors on the Trial Balance

a)  Total Dr bal $>$ Total Cr bal

|  | Dr | Cr |
|---|---|---|
| Case 1 | Overstate | Correct |
| Case 2 | Correct | Understate |
| Case 3 | Incorrect | Incorrect |

b)  Total Cr bal $>$ Total Dr bal

|  | Dr | Cr |
|---|---|---|
| Case 1 | Understate | Correct |
| Case 2 | Correct | Overstate |
| Case 3 | Incorrect | Incorrect |

## [05]  Effect of Errors on Profit

If errors are made in the Revenue Account, Expenses Account and Stock, the amount of gross profit and net profit will be affected.

| Account | Error | Effect on | |
| --- | --- | --- | --- |
| | | Gross Profit | Net Profit |
| Stock (Opening) | Understate | Overstate | Overstate |
| | Overstate | Understate | Understate |
| Stock (Closing) | Understate | Overstate | Overstate |
| | Overstate | Understate | Understate |
| Purchase | Understate | Understate | Understate |
| | Overstate | Overstate | Overstate |
| Purchase Returns | Understate | Understate | Understate |
| | Overstate | Overstate | Overstate |
| Sales | Understate | Overstate | Overstate |
| | Overstate | Understate | Understate |
| Sales Return | Understate | Understate | Understate |
| | Overstate | Overstate | Overstate |
| Trading Expense | Understate | Overstate | Overstate |
| | Overstate | Understate | Understate |
| Other Expenses | Understate | No Effect | Overstate |
| | Overstate | | Understate |
| Other Income | Understate | No Effect | Understate |
| | Overstate | | Overstate |

## Classwork 6

Office Furniture had been purchased for $3,600 on credit from Kowloon Co. on Jan 17, 20x6 that had not been recorded. The bookkeeper found this error on April 11, 20x7.

| Journal | | | | P.16 |
|---|---|---|---|---|
| Date | Particulars | F | Dr | Cr |
| 20x7 | | | $ | $ |
| Apr 11 | | | | |
| | | | | |
| | | | | |
| | Office Furniture purchased from | | | |
| | Kowloon Co. on Jan 17, 20x7 | | | |
| | had been omitted | | | |

## Classwork 7

$20,000 of goods sold on credit to Mr Chan had been entered in the account of Mr Chen. The Bookkeeper found this error on May 1, 20x7.

| Journal | | | | P.16 |
|---|---|---|---|---|
| Date | Particulars | F | Dr | Cr |
| 20x7 | | | $ | $ |
| May 1 | | | | |
| | | | | |
| | | | | |
| | | | | |
| | Sales entered in wrong cus- | | | |
| | tomer account | | | |

## Classwork 8

A sale of goods to Aberdeen Co. $690 had been entered in the books as $960 on Apr 11, 20x7. The Bookkeeper found this error on June, 20x7.

| | Journal | | | | P.17 |
|---|---|---|---|---|---|
| Date 20x7 Jun 7 | Particulars | | F | Dr $ | Cr $ |
| | Credit sales had been recorded overstating, now correct. | | | | |

## Classwork 9

$20,000 incurred to repair the Motor Vehicles in order to extent the life of that Motor Vehicle had been debited to Repair Expenses Account on Mar 19, 20x7. The bookkeeper found this error on July 26, 20x7.

| | Journal | | | | P.19 |
|---|---|---|---|---|---|
| Date 20x7 Jul 26 | Particulars | | F | Dr $ | Cr $ |
| | Cost of Motor Vehicles had been wrongly debited to Repair Expenses | | | | |

## Classwork 10

Cheque $3,000 received from Kwun Tong Co. had been entered twice in the Cash Book on Apr 16, 20x7. the bookkeeper found this error on Aug 18, 20x7.

| Journal | | | | | P.16 |
|---|---|---|---|---|---|
| Date | Particulars | F | Dr | Cr | |
| 20x7 | Correcting Entries | | $ | $ | |
| Aug 18 | | | | | |
| | Cheque received had been entered twice in the Cash Book | | | | |

## Classwork 11

Selling Expenses Account was overstated by $150 and the Sales Account was overstated by $150 too. The bookkeeper found these errors on September 16, 20x7.

| Journal | | | | | P.23 |
|---|---|---|---|---|---|
| Date | Particulars | F | Dr | Cr | |
| 20x7 | Correcting Entries | | $ | $ | |
| Sept 16 | | | | | |
| | Both the Selling and Telephone Expenses Account had been overstated by $100 | | | | |

## Classwork 12

Purchase of goods for cash $4,600 had been entered in the debit side of Cash account and posted to the Credit side of the Purchases Account. Error was found on 11 Nov 20x7.

| Journal | | | | | P.27 |
|---|---|---|---|---|---|
| Date<br>20x7<br>Nov 11 | Particulars | | F | Dr<br>$ | Cr<br>$ |
| | Purchased goods for cash<br>had been recorded in reverse | | | | |

## [06] Examinations

| |
|---|
| 2003 |
| 2004 |
| 2005 |

**Journal Voucher**

# Chapter 12

## Adjustment for Final Account

## Content

## [01] Introduction

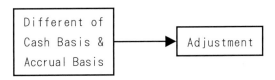

Adjustments:

◎Depreciation Expenses and Provision for Depreciation

◎Bad Debts and Provision for Bad Debts

◎Accrued Expenses

◎Prepaid Expenses

◎Accrued Revenue

◎Prepaid Revenue

## [02] Expenses

### Expenses

| | $ | | $ |
|---|---|---|---|
| Prepay b/d | X | Accrual b/d | X |
| Bank | X | Profit & Loss (Bal Fig) | X |
| Accrual c/d | X | Prepay b/d | X |
| | X | | X |

|  | Profit & Loss account | Balance Sheet |
|---|---|---|
| Expenses unpaid | Add | Current Liability |
| Expense prepaid | Less | Current Assets |

## Classwork 1

|  | 31/12/20X6 | 31/12/20X7 |
|---|---|---|
| Prepay Rent | $100 | $200 |
| Accrual Rent | $200 | $400 |
| Paid for Rent | $1,000 | $5,000 |

### Rent

## [02] Income

| Income | | | |
|---|---|---|---|
|  | $ |  | $ |
| In arrear b/d | X | In advance b/d | X |
| Profit & Loss (Bal Fig) | X | Bank | X |
| In advance c/d | X | In arrear c/d | X |
|  | X |  | X |

| | Profit & Loss account | Balance Sheet |
|---|---|---|
| Income in arrear | Add | Current Asset |
| Income in advance | Less | Current Liability |

## Classwork 2

| | 31/12/20x6 | 31/12/20x7 |
|---|---|---|
| Subscription in arrear | $1,000 | $1,500 |
| Subscription in advance | $200 | $400 |
| Bank | $5,000 | $7,000 |

## [03] Balance Sheet

| Balance Sheet (extract) | | | |
|---|---|---|---|
| Current Assets | $ | $ | $ |
| Stock | 100 | | |
| Debtors | 200 | | |
| Prepayment | | | |
| Subscription in arrear | | | |
| Bank | 1,000 | | |
| | | | |
| Less: Amount due within one year | | | |
| Creditor | 150 | | |
| Accrual Expenses | | | |
| Subscription in advance | | ___ | ___ |
| Net Current Assets | | | |

## [04] Adjustment on final account

The following balances were part of the trial balance of Maria on 31 December 20X7 :

| | Dr | Cr |
|---|---|---|
| | $ | $ |
| Stock at 1 January 20x6 | 2,070 | |
| Sales | | 22,000 |
| Purchase | 11,180 | |
| Rent | 650 | |
| Wages and salaries | 2,160 | |
| Insurance | 590 | |
| Bad debt | 270 | |
| Telephone | 300 | |
| General expenses | 180 | |

On 31 December 20X7 you ascertain that :

(a) The rent for four months of 2007 $160 had prepaid in 20x6.

(b) $290 owed for wages and salaries.

(c) Insurance had been prepaid $190.

(d) A telephone bill of $110 was owed.

(e) Stock on 31 December 20x7 was valued at $3,910.

Draw up Maria's trading and profit and loss account for the year ended 31 December 20X7 using the vertical style.

## STEP

(a) It is prepayment; less on the Trial Balance

(b) It is accrual expense; add on the trial balance

(c) It is prepayment.

(d) It is accrual

(e) Closing stock

[05] Examination

| 2003 | S2 | |
|------|----|----|
| 2004 | S2 | S3 |
| 2005 | S2 | |

# Chapter 13

## Capital and Revenue Expenditure

## Content

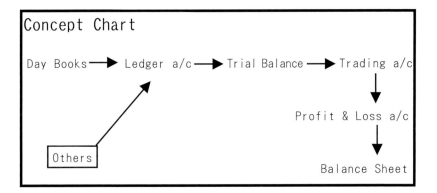

**Concept Chart**

Day Books ➤ Ledger a/c ➤ Trial Balance ➤ Trading a/c
↓
Profit & Loss a/c
↓
Others
Balance Sheet

## [01]    Introduction

| Expenditure | Details |
|---|---|
| Capital Expenditure | For buy Fixed Assets<br>For the use of new Fixed Assets<br>Upgrade the value of existing Fixed Assets |
| Revenue Expenditure | For business operation<br>Cannot upgrade the value of existing Fixed Assets |
| Joint Expenditure | Mixed with the above expenditures |

## [02]    Revenue

### a.Revenue expenditure

Revenue expenditure is incurred in keeping all fixed assets within the accounting period.

### b.Revenue receipts

Revenue receipts are the earnings from the operations of the business. Example are all goods sold and interest on investments.

## [03]  Capital

Capital expenditure means the purchase of fixed assets or expenditure which adds to the value of existing fixed assets. Examples are adding fitment to equipment.

### Example 1

| Expenditure | Type of Expenditure |
|---|---|
| a. Purchase of a motor van | Capital |
| b. Petrol costs for motor van | Revenue |
| c. Repairs to motor van | Revenue |
| d. Painting outside of new building | Capital |
| e. Five years later − repainting outside of building in (d) | Revenue |
| f. Extension to factory premises | Capital |
| g. Goods bought for resale | Revenue |
| h. Cleaning of office | Revenue |
| i. Wages and salaries $200,000 of which $20,000 paid to employees engaged on building an addition to the office premises | Capital $20,000 Revenue $180,000 |

## [04] Compare

| Capital Expenditure | Revenue Expenditure |
|---|---|
| It has long term nature | It has short term nature |
| It has an enduring influence on the profit-making capacity of the business | It is not a temporary influence on the profit-making capacity of the business |
| It is concerned with cost of purchasing fixed assets or adding to the value of an existing assets | It is not concerned with adding to the value of fixed assets, but represents the costs of running the business on day to day basis |
| In addition, it also included other costs necessary to get the fixed asset to operate | It is chargeable to the Trading, Profit & Loss account |
| It will result in increased figures for fixed assets in the Balance Sheet | |

## [04] Why it is so important to know their different?

The revenue expenditure would affect the profits. Over-state the revenue expenditure → net profit decrease.

## [05] Examination

| 2003 | |
|---|---|
| 2004 | S4 |
| 2005 | |
| 2006 | S2 |

# Chapter 14

## Correction of Error 1

## Content

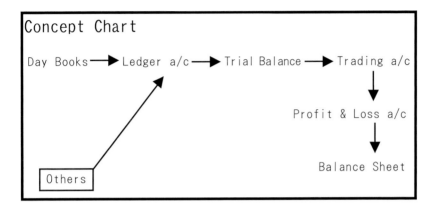

```
Concept Chart

Day Books ──▶ Ledger a/c ──▶ Trial Balance ──▶ Trading a/c
                       ↗                                │
                      ╱                                 ▼
                     ╱                         Profit & Loss a/c
                    ╱                                   │
            ┌──────┐                                    ▼
            │Others│                             Balance Sheet
            └──────┘
```

## [01]   Introduction

Errors in the books have:

1. do not show in Balance Sheet

2. show in the Trial Balance

When book-keepers find the mistakes, they must correct them by journal.

## [02]   Errors of Omission

The absence of both ledger accounts.

## Example 1

Office furniture purchased for $3,000 on credit from So Limited and has not been recorded.

|          | Dr               | Cr     |
|----------|------------------|--------|
| Right    | Office Furniture | So  Ltd |
| Actually | ____             | ____   |

## Correction

| Journal | | |
|---|---|---|
| | Dr | Cr |
| | $ | $ |
| Office furniture | | |
| So Ltd | | |
| | | |
| Correction of omission of capital purchase. | | |

## [03]  Errors of Commission

The wrong posting which is the same class.

### Example 2

A sale of goods $1,200 on credit to T So has been entered in the account of D So.

| | Dr | Cr |
|---|---|---|
| Right | Debtor − T So | Sales |
| Actually | Debtor − D So | Sales |

| Journal | | |
|---|---|---|
| | Dr | Cr |
| | $ | $ |
| Debtor − T So | 1,275 | |
| Debtor − D So | | 1,275 |
| | | |
| Sales entered in wrong customer account, now corrected. | | |

## [04]  Errors of Principle

The wrong posting which is the ~~different~~ class.

## Example 3

The payment of $1,000 for repairs has been wrongly post to motor van account.

|  | Dr | Cr |
|---|---|---|
| Right <br><br> Actually | Repairs | |

| Journal | | |
|---|---|---|
|  | Dr <br> $ | Cr <br> $ |
| Repairs | | |
|    Motor vehicles | | |
| Repairs wrongly debited to motor vehicle account, now corrected. | | |

## [05]  Errors of Original Entry

The original figures are posted incorrectly in both accounts.

## Example 4

Printing account is over-stated by $20 and sales account also is over-stated by $20.

|  | Dr | Cr |
|---|---|---|
| Right | Printing $100 | Sales $1000 |
| Actually |  |  |

Correction

| Journal | | |
|---|---|---|
|  | Dr $ | Cr $ |
| Sales |  |  |
| Over-statement in each of sales account and printing account, now corrected. | | |

## [06]  Errors of Compensations

Errors which posted in different sides in two accounts

## Example 5

A sale of goods $690 on credit to Lee has been entered in the books as $960.

|          | Dr | Cr |
|----------|----|----|
| Right    |    |    |
| Actually |    |    |

| Journal | | |
|---------|----|----|
|         | Dr<br>$ | Cr<br>$ |
| Sales understated, now<br>corrected. |  |  |

## [06]  Reversal of Entries

The account, which should be debit, has been credit, vice versa.

Example 6

Purchase of goods by cheque $1,000 has been entered wrong in both debit and credit sides.

|          | Dr | Cr |
|----------|----|----|
| Right    |    |    |
| Actually |    |    |

| Purchases | | | |
|---|---|---|---|
| | $ | | $ |
| Bank | | Bal b/d | 1,000 |

| Bank | | | |
|---|---|---|---|
| | $ | | $ |
| Bal b/d | 10,000 | Purchases | |

## [07] Examinations

| 2003 | | S3 |
|---|---|---|
| 2004 | | |
| 2005 | | |
| 2006 | S2 | |

# Chapter 15

## Correction of Error 2

### Content

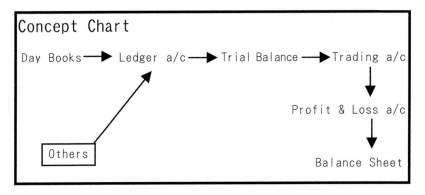

## Concept Chart

Day Books ➤ Ledger a/c ➤ Trial Balance ➤ Trading a/c

Others ➤ Ledger a/c

Trading a/c ➤ Profit & Loss a/c ➤ Balance Sheet

## [01]  Errors of Extraction

Example 1

Omission in the extraction of Sundry expenses $1,000

| Correction : | Dr | Sundry expenses | $1,000 |
|---|---|---|---|
| | Cr | Suspense | $1,000 |

## [02]  Errors in Posting

Advertising expenses $420 posted as $240

Example 2

| Correction : | Dr | Advertising ($420 − $240) | $180 |
|---|---|---|---|
| | Cr | Suspense | $180 |

## [03]  Errors of Incomplete Double Entry

The posting of cash payment for wages $700 has been omitted

Example 3

| Correction : | Dr | Wages | $700 |
|---|---|---|---|
| | Cr | Suspense | $700 |

## [04]   Errors in Addition

The total of the Returns inward books has been over-cast by $600

Example 4

| Correction : | Dr | Suspense | $600 |
| --- | --- | --- | --- |
| | Cr | Returns inwards | $600 |

## [05]   Errors in the Balancing of Accounts

Example 5

The balance of the Cash account has been under-stated to the extent of $500

| Correction : | Dr | Cash | $500 |
| --- | --- | --- | --- |
| | Cr | Suspense | $500 |

## [06]   Statement of corrected Net Profit

When we correct the errors, the net profit or loss will have been incorrectly. We also need to adjust the net profit or loss by the statement.

| Statement of adjusted the net profit | | |
| --- | --- | --- |
| | $ | $ |
| Net Profit before adjusted | | 100,000 |
| Add: (4) Overcast of Return Inwards | | |
| | | |
| Less:   (1) Omitted of Sundry Expense | | |
| (2) Error posting on Advertising | | |
| (3) Wage omitted | | |
| | | |
| Net Profit after adjusted | | |

## [07]  Revised Balance Sheet

As adjusting the entries have been made, correct the net profit for the period, the balance sheet should be revised to reflect the actual financial status of the business.

## [08]  Examination

| 2003 | | S4 |
|------|------|------|
| 2004 | S2 | |
| 2005 | | |

# Chapter 16

## Stock Valuation

## [01]  Introduction

If you buy a lot of good and some of them is not so famous. what would you do? You can sell it in a great discount.

Another case, after the "black rain", some of the goods have destroyed. You must spend some expenses before you sell it.

In the above two case, what should we do in book-keeping?

## [02]  Lower of cost or market value

Stock must be valued at the lower of its cost or net realizable value.

Saleable value — Expenses needed before completion of sale
= Net realizable value (NRV)

### Classwork 1

If the stock at cost valued $500 but the net realizable value is $400. Then how much would you be used as the value in the Trading a/c and Balance Sheet?

### Classwork 2

Stock Item          :PC01
Number in stock  :100
Cost price          :£150
Selling price      :Was £300 but reduced to £250 in February
                        20X7

Stock Item          :PC02
Number in stock  15
Cost price          :£165
Selling price      :£200 but 5 refrigerators required repairs
                        costing £60 each
Stock Item          :PC03

| | Cost | NRV |
|---|---|---|
| PC01 | | |
| PC02 | | |
| PC03 | | |

## [03]  Margin and Markup

Since the closing stock are not valued at the selling price, adjustments for goods sold and sale returns should therefore be converted into cost by deducting the profit loading from their sales values. Markup and margin are terms used to denote the profit of goods sold.

◎Markup = GP / COFS

◎Margin = GP / Selling Price

## Classwork 3

If cost price is $400,000; gross profit is $100,000, then what is his markup?

## Classwork 4

If cost price is $400,000; gross profit is $100,000, then what is his margin?

# Level 2

《LCCI Level 2 中級會計精讀備試天書》

# Content 目錄 / / / /

本書附送 LCCI 增值錦囊，請到以下網址下載：

http://www.systech.hk/lcci-note.rar

# Chapter 1
## Disposal of Fixed Assets

## [01] Introduction

Balance Sheet (extract)

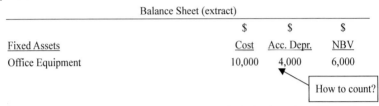

| Fixed Assets | $ Cost | $ Acc. Depr. | $ NBV |
|---|---|---|---|
| Office Equipment | 10,000 | 4,000 | 6,000 |

How to count?

Entities often buy assets intended for long-term use in the business that will provide a benefit for profit over a number of accounting periods.    Such assets are known as Fixed Assets.

Under the matching concept, it is necessary to charge a part of the cost of a fixed asset to EACH accounting period that is expected to benefit from its use.    Such a charge is known as Depreciation.

所有 classwork 的工作紙儲存於 CD 內，可以預先列印。

## [02] Methods of calculating depreciation charges

a.   *Straight Line Method - at an equal amount each year. (simply)*

| Depreciation per annum ← expense in P & L | = | Cost Price - Disposal Value No. of estimated year of use |
|---|---|---|

*Classwork 1*
If we bought a motor van for $20,000, thinking that we would keep it for 5 years and then sell it for $5,000.
[工作紙可於 CD 內下載]

b.   *Reducing balance method – based on the net book value of the asset*

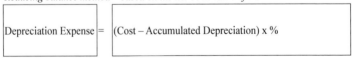

| Depreciation Expense | = | (Cost – Accumulated Depreciation) x % |
|---|---|---|

*Classwork 2*
If we bought a motor van for $20,000, thinking that we would keep it for 5 years and then will sell it for $5,000, depreciation rate is 20%.

c. *Revalued Method* 重估方法

| Bal b/d | - | Depreciation Expense | - | Disposal | = | Revalued Value |

---

*Classwork 3*
Beginning of the period $120,000          End of the period $100,000
Buy the new one $50,000                    Cost of old one $40,000

---

d) *Sum-of-the-year-digits*

Depreciation is apportioned according to the digit assigned in each financial year.

| Depreciation per annum | = | (Cost – Estimated Scrap Value) | x | No. of years of life remaining / Sum-of-the- digits |

---

*Classwork 4*
The original cost of a equipment is $50,000.   The expected useful life and the
estimated scrap value are 5 years and $5,000 respectively.

---

e) *Units of production / Machinery hours method*

Depreciation is calculated on the number of units of production in each financial
period out of the total available units.

| Depreciation per annum | = | (Cost – Estimated Scrap Value) | x | Production Unit in the year / Total Production Unit |

---

*Classwork 5*
The original cost of a equipment is $100,000.   The estimated scrap value and the
estimated number of production are $10,000 and 5,000 units respectively.

## [03] Double entry of Provision for Depreciation

We usually used to maintain each fixed assets at its cost while hold another ledger
account for depreciation.   This account is called "provision for depreciation".

*Which method of depreciation should be used?*
A different method might use for a certain asset.   For example, it is a fact that motor
vehicles lose value heavily early in their life and less so in later years.   The reducing
balance method therefore use appropriate because it indicates the way in which the
value of the asset is lost.

Dr: Profit & Loss a/c                    Cr: Provision for depreciation a/c

---

*Classwork 6*

A machine is bought for $10,000 on 1 Jan 20X5.   It is to be depreciated at the rate of 10% using the reducing balance method, prepare 2 years' account.

---

The fixed assets will be fully depreciated some years later.   If the fixed assets would still be operated after then, no depreciation would be calculated in following years and the net book value would be shown as "0".

## [04] Disposal

Upon the sale of a fixed asset, we will want to remove it from our ledger accounts. Some items which must appear in the disposal of fixed assets a/c are as follows:

| the original value of the asset at cost | + | the accumulated depreciation up to the date of sale | + | the sale prices of the asset. |
|---|---|---|---|---|

|  | DEBIT (Dr) | CREDIT (Cr) |
|---|---|---|
| [1] | Disposal of fixed asset account | Fixed asset a/c with the <u>cost</u> price. |
| [2] | Provision for depreciation account | Disposal of fixed asset account with the accumulated depreciation value |
| [3] | Cash book | Disposal of fixed asset account with the sale price of the asset. |
| [4] | The balance on the disposal account is the profit or loss on disposal and the corresponding double entry is recorded in the profit and loss account. | |

---

*Classwork 7:* Use the above data, sell the machinery $3,500 at 2 Jan 20X8

---

*Classwork 8:* Sell the machinery $9,000 at 2 Jan 20X8

---

|  | Dr | Cr |
|---|---|---|
| Profit on Disposal | Disposal | Profit & Loss |
| Loss on Disposal | Profit or Loss | Disposal |

Classwork 9
Richard Co. sold the machine at $2,000, which has bought at $10,000 and fully depreciation.    The depreciation rate is 20% by reducing balance method.

折舊是一課必考的課題,可以獨立一題或在 Final Accounts,所以考生必須熟讀計算方法及 Disposal 的格式。

= 本章完  End of Chapter =

# Chapter 2
## Provisions for doubtful debts & discounts

**[01] Introduction**

In this chapter you will learn the situation where there is some doubtful about the payment of the debt.

所有 classwork 的工作紙儲存於 CD 內，可以預先列印。

**[02] Why a business might create a provision for doubtful debts**

A provision for any bad and doubtful debt may be made when a firm thinks that there would be problems in recovering a debt.

● Prudence Concept - can avoid overstating profits and debtor

● Matching Concept – can relate any loss for the period with the related sales in the same period

**Provision for doubtful debts** shows the estimated amounts for debtors, who will be unable to pay their accounts, at the balance sheet date.

Using current period outstanding debtors' balance, a company would forecast the new provision of doubtful debts for the coming financial period.

Provision for doubtful debts

= | Debtors after bad debts (net debtors) | x | rate (%) |

**[03] The posting for provision for doubtful debts**

The provision reduces the value of debtors in an attempt to give their real worth to the business.　It is a contra-asset item charged as an expense.

The double entry will be: (10 →12)

| Dr | Cr |
|---|---|
| Profit and Loss Account | Provision for Bad Debts account |

*In Year 1, The amount of debtor is $100,000 and the rate of provision for doubtful debt is 10% of outstanding debts.　In Year 2, the amount of debtor is $120,000 and the rate is remaining unchanged.*

### Provision for Doubtful Debts

| Year 2 | | $ | Year 2 | | $ |
|---|---|---|---|---|---|
| | | | Jan 1 | Bal b/d [1] | 10,000 |
| | | | | (100,000 x 10%) | |
| Dec 31 | Bal c/d [2] | 12,000 | Dec 31 | Profit & Loss (bal) | 2,000 |
| | (120,000 x 10%) | | | | |
| | | 12,000 | | | 12,000 |
| | | | Year 3 | | |
| | | | Jan 1 | Bal b/d | 12,000 |

---

*Classwork 1*

In Dec 20X7, our provision for doubtful debt increases from $10,000 to $12,000.

---

Should the amount required for the provision have to be reduced, the provision account must reduce its credit balance. The corresponding credit entry will be made in the profit and loss account as <u>income</u>.

The double entry is: (12→10)

| Dr | Cr |
|---|---|
| Provision for Bad Debts account | Profit & Loss account |

*In Year 2, The amount of debtor is $120,000 and the rate of provision for doubtful debt is 10% of outstanding debts. In Year 3, the amount of debtor is $110,000 and the rate is remaining unchanged.*

### Provision for Doubtful Debts

| Year 2 | | $ | Year 2 | | $ |
|---|---|---|---|---|---|
| | | | Jan 1 | Bal b/d | 10,000 |
| | (120,000 x 10%) | | | (100,000 x 10%) | |
| Dec 31 | Bal c/d | 12,000 | Dec 31 | Profit & Loss (bal) | 2,000 |
| | | 12,000 | | | 12,000 |
| Year 3 | | | Year 3 | | |
| Dec 31 | Profit & Loss (bal) | 1,000 | Jan 1 | Bal b/d [1] | 12,000 |
| | Bal c/d [2] (110,000 x 10%) | 11,000 | | | |
| | | 12,000 | | | 12,000 |
| | | | Year 4 | | |
| | | | Jan 1 | Bal b/d | 11,000 |

*Classwork 2*
In Dec 20X7, our provision for doubtful debt decreases from $12,000 to $9,000.

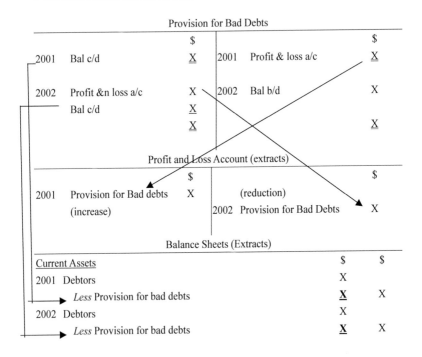

Provision for Bad Debts

| | | $ | | | $ |
|---|---|---|---|---|---|
| 2001 | Bal c/d | X | 2001 | Profit & loss a/c | X |
| 2002 | Profit &n loss a/c | X | 2002 | Bal b/d | X |
| | Bal c/d | X | | | |
| | | X | | | X |

Profit and Loss Account (extracts)

| | | $ | | | $ |
|---|---|---|---|---|---|
| 2001 | Provision for Bad debts (increase) | X | | (reduction) | |
| | | | 2002 | Provision for Bad Debts | X |

Balance Sheets (Extracts)

| Current Assets | | $ | $ |
|---|---|---|---|
| 2001 | Debtors | X | |
| | Less Provision for bad debts | X | X |
| 2002 | Debtors | X | |
| | Less Provision for bad debts | X | X |

*Classwork 3*

| Year | Debtor Balance ($) | Bad Debt ($) | Net Debtor ($) | Doubtful debt rate |
|---|---|---|---|---|
| 1 | 10,000 | 100 | | 5% |
| 2 | 15,000 | 100 | | 5% |
| 3 | 20,000 | 100 | | 1% |

Prepare the profit & loss account and balance sheet.

**[04] Different between bad debt and provision for doubtful debt**

| | Bad Debt | Provision for bad debt |
|---|---|---|
| Types | Expenses | Expenses / Incomes |
| LCCI | Level 1 | Level 2 & 3 |
| Placing | PL | PL / CA |
| Accuracy | Known with certainly | Estimate of possible losses |

7

## [05] Provision for Discounts

In the business environment, it is estimation of the discounts allowed which occurred in coming period, based on the existing level of debtors balance. The amount of the Provision for Discounts Allowed is calculated as a certain percentage of net debtor balance which after the deduction provision for bad and doubtful debts at the end of each financial period.

*Classwork 4*

| Description | 2005 ($) | 2006 ($) | 2007 ($) |
|---|---|---|---|
| Debtors bal c/d | 2,000 | 3,000 | 2,000 |
| Bad Debts | 500 | 400 | 500 |
| Net Debtors | 1,500 | 2,600 | 1,500 |
| Provision for Bad Debts (%) | 5 | 10 | 15 |
| Provision for Bad Debts ($) | (75) | (260) | (225) |
| | | | |
| Provision for discount allowed (%) | 4 | 3 | 4 |
| Provision for discount allowed ($) | (57) | (70.2) | (51) |
| | | | |

*Classwork 5*
Debtor $100,000, provision for doubtful debts $2,000, discount for provision is 5%, bad debt $10,000. Prepare the balance sheet.

以往 Provision for doubtful debt 是屬於 Level 1 的課題,現改為 Level 2,本人覺得其中一個可能性是因為很多 Level 1 的考生未能掌握其中要點,尤其是 double entry 的過程,考生必須熟習其中過程,因為此題目是經常混入 Balance Sheet 內的,是屬於大熱的課題。

= 本章完 End of Chapter =

# Chapter 3
## Income & Expenditure a/c

## [01] Introduction

Club and society provides services for member. It is not for profit-making.

所有 classwork 的工作紙儲存於 CD 內，可以預先列印。

## [02] Types of Organisations

Private Enterprise (profit making)

Sole Proprietor + Partnership + Limited Company + Franchising + Co-operative

Public Enterprise (Non-profit making)

Public Corporations + Government Departments + Local Authorities

Non-profit Making Organisations:

Club, Associations, Parties, Hospitals, Schools, Charitable bodies

## [03] Receipts and Payments Account

| Profit Making Firm | Non-profit Making Organization |
|---|---|
| Cash Book | Receipts and Payments Account |

### Receipts and Payments a/c

| | $ | | $ |
|---|---|---|---|
| Bal b/d | 200 | Bar wages | 700 |
| Subscriptions | 1000 | Sports stadium expenses | 200 |
| Rent received | 100 | Bal c/d | 400 |
| | 1300 | | 1300 |

## [04] Income and Expenditure Accounts

*Income and expenditure account* is an account for a non-profit making organization to find the surplus or deficit made during a period.

| Profit Making Firm | Non-profit Making Organization |
|---|---|
| 1. Trading account | 1. Trading account |
|    ●   sales |    ●   Bar taking |
|    ●   gross profit / loss |    ●   Bar profit / loss |
| 2. Profit and Loss account | 2. Income & Expenditure a/c |
| 3. Net profit | 3. Surplus |
| 4. Net loss | 4. Deficit |

## [05] Accumulated Fund

Accumulated Fund = Assets - Liabilities

| Profit Making Firm | Non-profit Making Organization |
|---|---|
| 1. Capital | 1. Accumulated Fund |
| 2. Drawing | 2. Not available |

*Classwork 1*

Sport Equipment $8,000, Club Premises $22,000, Furniture & Fittings $5,000, Prepaid Rates $200, Bank $2,000, Creditor $1,000, $ Bar Stock $2,000
Use the above information to prepare the statement of affairs.

## [06] Subscriptions (Income)

The major source of incomes of a non-profit making organisation is the subscriptions of members.

Subscriptions

| 20X7 | | | $ | 20X7 | | | $ |
|---|---|---|---|---|---|---|---|
| Jan 1 | In arrear b/d | (1) | 300 | Jan 1 | In advance b/d | (2) | 100 |
| Dec 31 | I & E (Bal fig) | | *7000 | Dec 31 | Bank | (3) | 7,000 |
| | In advance c/d | (3) | 100 | | In arrear c/d | (5) | 300 |
| | | | 7,400 | | | | 7,400 |
| 20X8 | | | | 20X8 | | | |
| Jan 1 | In arrear b/d | (5) | 300 | Jan 1 | In advance b/d | (4) | 100 |

*Classwork 2*

Received during the year 550
Not yet receive at year beginning 350            at year end 200
Receive in advance at year beginning 600         at year end 300

## [07] Donations (Income)

Any *donations* received are shown as income in the year that they are received. It can be from members or non-members.

The double entry of donations and entrance fees

| Description | Double Entry |
|---|---|
| Donation received or entrance fee received | Dr: Cash    Cr: Donation |

**[08] Receipts from specific activity**

Any income and expenditure in a specific activity should be matched and shown in the Income & Expenditure a/c to reflect the profit of loss on the specific activity

| Profit on activity will be treated as income | Actual Income > Actual Expense |
|---|---|
| Loss on activity will be treated as expenditure just like profit / loss on disposal. | Actual Income < Actual Expense |

**[09] Expenses Items**

- Rent & Rates for organisation's premises
- Staff Salaries and Wages
- Maintenance Cost of Premises
- Other Running Expenses
- Loss on Fund Raising Activities

**[10] Format of Final Accounts**

Income & Expenditure a/c of a non-profit making organisation is used to reflect the actual income and expenditure in each financial year, whether the amounts of them are received or paid.

a) Format of Income & Expenditure a/c

Richard Club

Income & Expenditure a/c for the year ended 31 Dec 2007

| *Incomes* | $ | $ |
|---|---|---|
| Subscription | 100 | |
| Life Membership Fee | 200 | |
| Bar Profit | 100 | |
| Entrance Fee | 200 | 600 |
| | | |
| *Expenditures* | | |
| Salaries | 200 | |
| Rent | 100 | 300 |
| Surplus | | 300 |

Incomes > Expenditure → Surplus

Expenditure > Incomes → Deficit

b) The format of the Balance Sheet of both will be identical. "Accumulated Fund (AF)" will replace the item "Capital" in the organisation.

| Opening AF | = | Opening Total Assets | - | Opening Total Liabilities |

## [11] Bar Trading Account

A separate trading account is to be prepared for profit section such as bar, restaurant, refreshment, where is made for one kind of income.

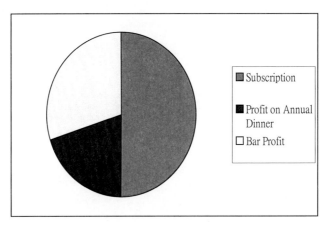

Incomes of the club and society

*Classwork 3*

Trial Balance as at 31 December 20X7

|  | Dr $. | Cr.$ |
|---|---|---|
| Equipment | 8,000 | |
| Premises | 29,000 | |
| Subscriptions | | 6,000 |
| Wages | 4,000 | |
| Furniture and fittings | 5,000 | |
| Rates and Rent | 1,900 | |
| Sundry expenses | 600 | |
| Accumulated fund 1 January 20X7 | | |
| Donations | | 300 |
| Telephone ,Postage & Printing | 500 | |
| Ban k | 2,000 | |
| Bar purchases | 9,500 | |
| Creditors | | 1,000 |
| Bar sales | | 14,000 |

Bar stocks 1 January 20X7                                      2,000

---

*Classwork 4*

The following information is also available: (i) Bar stocks at 31 December 20X7 $2,000.   (ii) Provision for depreciation: sports equipment $1,700; Furniture and fittings $1,200.

---

Surplus =                              >
Deficit =                              <

## [12] Different between Income & Expenditure a/c and Receipt & Payment a/c

|  | Receipts & Payment a/c | Income & Expenditure a/c |
|---|---|---|
| Nature | Record the cash in and out briefly | Calculate the surplus and deficit |
| Entry | Cash Book | Profit & Loss a/c |

= 本章完  End of Chapter =

## Chapter 4
## Stock Valuation

### [01] Introduction

S K Ltd purchased and sold gold to earn profit.   The following is its trading during the month of Jan 20X8.

| Gold bought (20X8) | Gold sold (20X8) |
|---|---|
| Jan 1    10 at $200 @ | Jan 17   8 at $360@ |
| Jan 12   10 at $240@ | Jan 29   24 at $400@ |
| Jan 28   20 at $250@ | |

What should be the closing stock at Jan 20X8?

**Problem** in making decision: Gold bought in different prices

**Solution**: Determine the ways to calculate the value of sock

Stock should not be valued at selling price as to do so would anticipate profit which has not yet been earned by the business.

所有 classwork 的工作紙儲存於 CD 內，可以預先列印。

### [02] What is "stock"?

- for sale in the operating activities of the business; or
- In the process of production for sales such as work in progress; or
- In the form of materials for production such as raw materials.

### [03] Net Realisable Values (NRV)

| Saleable value | - | Expenses needed before completion of sales | = | NRV |
|---|---|---|---|---|

### [04] Measurement of Stock

Stock must be valued at the lower of its cost or net realizable value.

*Classwork 1*

| Article | Different Categories | Cost |
|---|---|---|
| 1 | A | 100 |
| 2 | A | 120 |
| 3 | A | 300 |
| 4 | B | 180 |

| 5 | B | 150 |
|---|---|-----|
| 6 | B | 260 |
| 7 | C | 410 |
| 8 | C | 360 |
| 9 | C | 420 |

| Category | Cost | Stock valuation |
|----------|------|-----------------|
| A | $(100+120+300) = $520 | $100 x 3 = $300 |
| B | | |
| C | | |
| Total | | |

The value of closing stock can be calculated under perpetual or periodic stock system.   However, they are still subject to the following adjustments
- Lower of cost or market value
- Movement of stock

## [05]   Methods of stock valuation

a)   Cost of closing stock

$$\text{Cost of closing stock} = \text{Costs of purchase} + \text{Costs of Conversion} + \text{Other Costs}$$

Costs of Purchase
- Invoice price, Import duties, Trade discounts, other Taxes
- Transportation for materials, service and finished goods
- Other costs directly attributable to the buying

Cost of Conversion  加工成本
- Cost directly related to the units of production
- Production cost which incurred in converting materials into finished goods

Other Costs

They are included in the cost of stock in bringing stock to their present location and condition. For example, the costs of designing products for specific customers in the cost of stock in South Africa.

b) Examples of costs excluded from the cost of stock
- Abnormal (不可預期的) amounts of wasted materials, labour or other related production costs.
- Storage costs (treat as administrative expense in Profit & Loss a/c)
- Administrative Overheads (post in Profit & Loss a/c)
- Selling and transportation costs (treat as selling and distribution expense in Profit & Loss a/c)

c) First in, First out method (FIFO)

**FIFO** is a method by which the first goods to be received are aid to be the first to be sold.

| | Received | Issued | Stock after each transaction | | |
|---|---|---|---|---|---|
| 20X7 | | | | $ | $ |
| January | 10 at $300 each | | 10 at $300 each | | 3000 |
| April | 10 at $340 each | | 10 at $300 each<br>10 at $340 each | 3000<br>3400 | 6400 |
| May | | 8 x | 2 at ____ each<br>10 at ____ each | | 4000 |
| October | 20 at $400 each | | 2 at ____ each<br>10 at ____ each<br>20 at ____ each | | 1,2000 |
| November | | 2  x<br>10 x<br>12 x<br>_24_ | 8 at ____ each | | |

| What is the Cost of Good Sold? |
| --- |
|  |

d) Last in, First out method (LIFO)

**LIFO** is a method by which the goods sold are said to have come from the last batch of goods to be received.

| | Received | Issued | Stock after each transaction | | |
| --- | --- | --- | --- | --- | --- |
| 20X7 | | | | $ | $ |
| January | 10 x $300 | | 10 at $300 each | | 3000 |
| April | 10 at $340 | | 10 at $300 each | 3000 | |
| | each | | 10 at $340 each | 3400 | 6400 |
| May | | 8 x | 10 at ____ each | | |
| | | | 2 at ____ each | | |
| October | 20 at $400 | | 10 at ____ each | | |
| | each | | 2 at ____ each | | |
| | | | 20 at ____ each | | |
| November | | 20 x | | | |
| | | 2 x | | | |
| | | 2 x | 8 at ____ each | | |
| | | <u>24</u> | | | |

*Classwork 3*

| What is the Cost of Good Sold? |
| --- |
|  |

If we use the LIFO method, we assume that issues are always taken from the most recent deliveries.

e) Weighted average method (WAVO)

WAVO means the cost of units are averaged, and units used are charged at average cost.

| | Received | Issued | Total Value | Number | Average |
| --- | --- | --- | --- | --- | --- |
| | | | $ | | |
| 20X7 | | | | | |
| January | 10 at $30 | | 10 x $300 | 10 | |

| April | 10 at $3400 | | | 20 | |
| May | | 8 at $340 | | 12 | |
| October | 20 at $400 | | | 320 | |
| November | | 24 at $370 | | 80 | |

*Classwork 4*

What is the Cost of Good Sold?

**[06]  Compare the different effects on the reported profit of the business**

- With the WAVO, the items are issued at the average cost per unit, and the issue price is determined by dividing the total value by the number of units in stock

- This will tend to smooth out price fluctuations and the closing stock valuation will fall between that resulting from the FIFO and LIFO methods.

- In times of rising prices, the cost of good sold figure for the WAVO will higher than FIFO but lower than LIFO

- The profit will be lower than that calculated under FIFO but higher than that calculated under LIFO

*Classwork 5*

| | FIFO | | | LIFO | | | WAVO | |
|---|---|---|---|---|---|---|---|---|
| | $ | $ | | $ | $ | | $ | $ |
| Sales | | 50,000 | | | 50,000 | | | 50,000 |
| Less: COGS | | | | | | | | |
| Gross Profit | | | | | | | | |

**[07]  Loss of Stock**

It may be required to calculate the loss of stock at a particular date sue to goods stolen by thieves or destroyed by fire.   Similar adjustments as made in the above example are needed to arrive at the amount of the loss.

=本章完  End of Chapter =

| | Chapter 5 |
|---|---|
| | **Partnership – Format** |

打劫都會因為分錢而出問題，試問合夥做生意時，如何可以避免此問題?

## [01] Introduction

| | Sole Trader Business | Partnership[1] |
|---|---|---|
| Owner / Owners | One | Two or more partners |
| Appropriation of net Profit | Not applicable as all the profit belongs to the sole | Shared by partners |
| Final Accounts | Trading a/c Profit & Loss a/c Balance Sheet | Trading a/c Profit & Loss a/c Appropriation a/c Balance Sheet |

---

[1] Partnership Act 1890 defines a partnership as "the relationship which subsists between persons carrying on a business in common with a view to profit"

(a) The need for partnership

Partnerships are firms in which two or more people are working together to try to make profits.

(b) Nature of a partnership

Each partner, except for the limited partners, is liable to pay his share of partnership debts.

(c) Types of Partners

| *General Partners* |
| *(at lease 1 in partnership)* |

| *Sleeping / Silent Partners* |

**[02]   Partnership Agreements**

Written partnership agreements are preferable to mere oral agreements.

Underlying:

| Capital Contribution | Profit / Loss sharing ratio | Interest on Capital |

| Interest on Drawings | Salaries |

| Sharing of Loss on realisation | Other special terms and conditions |

(a) Profit or Loss sharing ratios

Profits or losses do not have to be shared in ratio to capital. It set in the partnership agreement such as Peter : Mary is 2:1.

(b) Interest on Capital

Interest on Capital can compensate for unequal investment of partners.

| Capital a/c balance amount | x | Rate (%) | = | Interest on Capital |

*Classwork 1*

| Capital – A $20,000; B $15,000, Net Profit $12,000, Profit & Loss Sharing A to B : 1:1, Interest on Capital is 5%. What is the amount of total interest on capital? |

所有 classwork 的工作紙儲存於 CD 內，可以預先列印。

(c) Interest on Current

Interest on Current can compensate for unequal current being contributed.  It can be positive or negative.

| Current a/c balance amount | x | Rate (%) | = | Interest on Current |
|---|---|---|---|---|

*Classwork 2*

Example: Current a/c – A ($20,000); B $15,000; Interest on Current is 5%

(d) Salaries to partner

A salary can compensate a partner for extra duties or responsibilities.

*Classwork 3*

Lau & Cheung was the partners.   Lau was entitled to a monthly salary of $2,400. The salary of Cheung was half that Lau from 1 Apr inwards, they receive new salary instead the old one.   Lau monthly rate was increased by 10% upon the formation of the company while, Cheung's monthly rate was increased by 20% with effect from 1 July.   What is their Partner's Salary?

(e) Interest on drawing

Interest on Drawing encourages partners to avoid excessive drawings.

| Total Drawings | = | Cash Drawings | + | Stock Drawings |
|---|---|---|---|---|

| Total Drawings | x | Rate (%) | = | Interest on Drawings |
|---|---|---|---|---|

*Classwork 4*

Example: Capital – A $20,000; B $15,000, Net Profit $50,000, Profit & Loss Sharing A to B : 1:1, Interest on Capital is 5%, Drawings – A:$2,000, B $4,000, Interest on Drawings is 1%, Partner Salary of both is $1000

| Double entry | Dr | Cr |
|---|---|---|
| Interest on Capital<br>Partner Salary<br>Net Profit Sharing | Appropriation a/c | Partners' Current a/c |

21

| Drawings | Partners' Current a/c | Cash |
|---|---|---|
| Interest on Drawings | Partners' Current a/c | Appropriation a/c |

## [03] The final accounts

Profit and Loss a/c

- would be same as the sole traders'
- however, a Appropriation a/c will be opened to show the distribution of profits or losses among partners.

Richard Ltd

Appropriation a/c for the year ended 31 December 20X7

|  | $ | $ | $ |
|---|---|---|---|
| Net Profit before Appropriation |  |  | 10,000 |
| Less: Appropriation |  |  |  |
| Profit on Capital |  |  |  |
| ~ A | 100 |  |  |
| ~ B | 200 |  |  |
| ~ C | 100 | 400 |  |
| Partner Salary |  |  |  |
| ~ A | 100 |  |  |
| ~ B | 100 |  |  |
| ~ C | 100 | 300 |  |
| Interest on Current |  |  |  |
| ~ A | (200) |  |  |
| ~ B | 200 |  |  |
| ~ C | 100 | 300 |  |
| Interest on Drawings |  |  |  |
| ~ A | 200 |  |  |
| ~ B | 300 |  |  |
| ~ C | 100 | (600) | 400 |
| Net Profit after Appropriation |  |  | 9,600 |

From Profit & Loss a/c

Profit Sharing:

|  |  |
|---|---|
| ~ A | 3,200 |
| ~ B | 3,200 |
| ~ C | 3,200 |
|  | 9,600 |

Balance Sheet
- also the same as the sole trader's in the part of Assets, Liabilities except Capital

Richard Ltd
Balance Sheet (extract) as at 31 December 20X7

| Financed by: | A | B | C | |
|---|---|---|---|---|
| Capital a/c | 10,000 | 10,000 | 10,000 | 30,000 |
| Current a/c | 8,000 | (1,000) | 5,000 | 12,000 |
| | 18,000 | 9,000 | 15,000 | 42,000 |

| Long Term Liabilities | | |
|---|---|---|
| Loan | | 20,000 |
| | | 62,000 |

## [04] Capital Accounts and Current Account

The Capital account remain fixed the amount introduced as capitals. The other items including interest on capital, salaries, share of profits and interest on drawing are entered in the partners' current accounts via the profit and loss appropriation account.

Capital a/c

| | A $ | B $ | C $ | | A $ | B $ | C $ |
|---|---|---|---|---|---|---|---|
| | | | | Bal b/d | | | |
| Bal c/d | | | | Cash | | | |
| | | | | | | | |

Current a/c

| | A $ | B $ | C $ | | A $ | B $ | C $ |
|---|---|---|---|---|---|---|---|
| Bal b/d | | | | Bal b/d | | | |
| Interest on Current | | | | Interest on Capital | | | |
| Drawings | | | | Interest on Current | | | |
| Interest on Drawings | | | | Partner Salary | | | |
| Net Loss | | | | Net Profit | | | |
| Bal c/d | | | | Bal c/d | | | |
| | | | | | | | |

= 本章完 End of Chapter =

**[01] Introduction**

*Goodwill* represents the **reputation** and **intangible** value of an **established** going concern.

● An intangible assets

所有 classwork 的工作紙儲存於 CD 內，可以預先列印。

**[02] Goodwill exists because of:**

1. a good reputation – brand name
2. experienced, efficient & reliable employees – academic level
3. superior management team
4. good location
5. large, number of regular, customers
6. Going Concern – is it has profit growth?
7. Experience on the industry
8. The Possession of Favorable Contracts

**[03] Circumstances calling for the ascertainment of the value of goodwill are:**

1. Admission or withdrawal of Partners
2. Changes in Profit and Loss Sharing Ratio
3. Retirement or death of partner(s)
4. Amalgamation (combination) or transfer of business

*Classwork 1*

| |
|---|
| What is goodwill? Under what situation that will be record in a partnership? How to post in Balance Sheet? |

**[04] Nature of Goodwill**

| Total Price | - | Value of Assets | = | Goodwill |
|---|---|---|---|---|

*Classwork 2*

| You want to sell the business and assets and liabilities are listed: | | | |
|---|---|---|---|
| Buildings | 225,000 | Long Term Liabilities | 100,000 |
| Machinery | 75,000 | Creditors | 30,000 |
| Debtors | 60,000 | Bank Overdraft | 45,000 |
| Stock | 40,000 | | |

Mr. So paid $300,000 for the whole business, what is the amount of goodwill?

**[05]  Types of Goodwill**

| Positive Goodwill | Negative Goodwill |
|---|---|
| The amount of the purchase consideration minus the fair value of the separate net assets of the business | Goodwill does not necessary exist in a business if it had a bad reputation, in efficient labour force, etc. |

**[06]  Valuation**

*Goodwill* is concerned with future profits. However, there is no single way of calculating goodwill to which everyone can agree but agreement reached between buyer and seller.

(i)  Average Sales Method

| Average Yearly Sale for the past few years | x | Agreed Figure |
|---|---|---|

- Usually for retail businesses

*Classwork 3*

It is agreed that the goodwill should be 3 times the average yearly sales for the past three years.  Calculate the value of goodwill:

Sales for the past three years:

| 20X5 | $3,650,000 |
|---|---|
| 20X6 | $5,000,000 |
| 20X7 | $6,350,000 |

(ii)  Annual Fees Method

- For professional firms
- Based on gross annual income from fees, before charging expenses

*Classwork 4*

A CPA firm is selling its business.  It is asking a figure for goodwill, which is 2.5 times the average annual fees, received for the last two years.  Calculate the goodwill:

| 20X6 | $1,800,000 |
|---|---|
| 20X7 | $2,200,000 |

(iii) Average Net Annual Profits Method

| Average Net Profit for a no. of years | x | a stated amount |
|---|---|---|

*Classwork 5*

Suppose goodwill is taken to be 4 times the average net profits for the past years. Calculate the goodwill.

| | |
|---|---|
| 20X5 | $6,200,000 |
| 20X6 | $6,900,000 |
| 20X7 | $7,900,000 |

**[07]  Book- keeping Entries**

(i)  Goodwill Account Opened – stated in the Balance Sheet as "Intangible Fixed Assets"

| Descriptions | Book-keeping Entries |
|---|---|
| Shared by EXISTING partners (the OLD profit sharing ratio) | **Dr.**  Goodwill<br>    **Cr.** Partners' Capital Accounts |

(ii)  Goodwill Account **NOT** Opened

| Descriptions | Book-keeping Entries |
|---|---|
| Shared by EXISTING partners (the OLD profit sharing ratio) | **Dr.**  Goodwill<br>    **Cr.** Partners' Capital Accounts |

| Shared by partners (the NEW profit sharing ratio) | **Dr.**  Partners' Capital Accounts<br>    **Cr.** Goodwill |

| |
|---|
| Goodwill account is closed. |

**[08]  Goodwill on Admission of partners**

*Classwork 6*

R is admitted by bring in $4,000 cash.

Goodwill is agreed to be $10,000

Old sharing ratio: L(2), M(5), S(3)

New sharing Ratio: L(1), M(5), S(3), R(1)

(a) Not to open a goodwill account – New partner pays his share of goodwill in cash into the business. Old partners share the $ by increasing their capital, i.e., goodwill adjustment.

(b) Not to open a goodwill account – New partner pays cash privately to old partners for his share of goodwill

## [09] Goodwill on withdrawal of partner

*Classwork 7*

A,B,C are partners share profits equally.
C withdraws from the partnership
Capital opening balances; $4,000 each
Goodwill is agreed to be $3,000 when C withdraws

(a) When there is no Goodwill Account in the partnership:

| Dr | Cr |
|---|---|
| Goodwill a/c | Capital a/c (old ratios) |
| Capital – partners gained | Capital – partners lost |

Please prepare the accounts?
When C leaves the partnership, $_____ will be paid to him.

(b) If there is a Goodwill Account in the partnership:
(i) Existing goodwill amount is the same ->
● If a goodwill a/c exits with correct valuation of goodwill entered it, no further action
(ii) Goodwill is undervalued: e.g. goodwill already exist: $900
Please prepare the accounts?

(c) Goodwill is overvalued:
e.g. Goodwill already exist: $3,600
Please prepare the accounts?

## [10] Goodwill on withdrawal and admission

Peter, John and May are partners sharing P & L in the ratio 1:2:3 respectively. May retired and June was admitted as a partner. The new agreed P & L sharing ration of Peter, John and June are 2:3:5 respectively. Goodwill was valued at $1,200.00.

| Goodwill Account Opened | | |
|---|---|---|
| Dr.     Goodwill | $1,200 | |
|     Cr.     Capital Accounts | | |
|         - Peter (1/6) | | $200 |
|         - John (2/6) | | $400 |
|         - May (3/6) | | $600 |

| Goodwill Account NOT Opened | | |
|---|---|---|
| Dr.     Goodwill | $1,200 | |
|     Cr.     Capital Accounts | | |
|         - Peter (1/6) | | $200 |
|         - John (2/6) | | $400 |
|         - May (3/6) | | $600 |
| Dr.     Capital Accounts | | |
|         - Peter (2/10) | $240 | |
|         - John (3/10) | $360 | |
|         - June (5/10) | $600 | |
|     Cr.     Goodwill | | $1,200 |

**Statement of Goodwill Adjustments Between Partners**

| | | Peter | John | May | June | Total |
|---|---|---|---|---|---|---|
| | | $ | $ | $ | $ | $ |
| Goodwill was shared | 1:2:3:- | 200 | 400 | 600 | - | 1,200 |
| Goodwill was written off | 2:3:-:5 | (240) | (360) | - | (600) | 1,200 |
| Net Adjustments | | (40) | 40 | 600 | (600) | - |

= 本章完 End of Chapter =

同學可於 CD 內開啓本課的 POWERPOINT，內有詳細的中文解釋如何處理商譽。

## [01] Introduction

The assets and liabilities of a business may be revalued:

(1) In admission or retirement of a partner, the assets and liabilities may be revalued to reflect the fair value of the business.

(2) The partners change profit and loss sharing ratio

(3) The business has be taken over by a limited company

(4) Two or more partnerships are amalgamated

## [02] Need for Revaluation

- The new partner would benefit from increase in value of assets before he joins the firm without paying anything
- The new partners would loss from decrease in value of assets with no revaluation takes place
- The assets of the partnership business should be revalued at the date of change so that any revaluation gain or loss of revaluation may be distributed to the partners before the change

## [03] Book-keeping Entries

| Description | Book-keeping Entries | | |
|---|---|---|---|
| Transfer the reduction in the value of assets to Revaluation Account | **Dr** | Revaluation Account | |
| | | **Cr** | Assets Account |
| Transfer the increase in the value of assets to Revaluation Account | **Dr** | Assets Account | |
| | | **Cr** | Revaluation Account |
| Share Profit on Revaluation (Profit Sharing Ratio) | **Dr** | Revaluation Account | |
| | | **Cr** | Partners' Capital Accounts |
| Share Loss on Revaluation (Loss Sharing Ratio) | **Dr** | Partners' Capital Accounts | |
| | | **Cr** | Revaluation Account |

Revaluation Account

| REDUCTION In the value of assets | INCREASE In the value of assets |
|---|---|
| PROFIT ON REVALUATION | LOSS ON REVALUATION |

*Classwork 1*

Land $5,000 → $10,000

所有 classwork 的工作紙儲存於 CD 內，可以預先列印。

*Classwork 2*

| |
|---|
| Car $15,000 → $11,000 |

**[04]    Profit or Loss on Revaluation**

IF: New total valuation of assets > Old total valuation of assets = Profit on Revaluation

IF: New total valuation of assets < Old total valuation of assets = Loss on Revaluation

*Classwork 3*

| Items | Dr | Cr |
|---|---|---|
| Showing gain on revaluation | | |
| Showing loss on revaluation | | |
| Showing an increase in total valuation of assets | | |
| Showing a fall in total valuation of assets | | |

*Classwork 4*

W and Y share profit and losses in ratio 2/3, 1/3 at 31 December 20X7.    From 1 January 20X8, the ratio would be shared equally.

| Balance Sheet as at 31 December 20X7 | | | |
|---|---|---|---|
| Premises | 65,000 | Capital W | 70,000 |
| Fixtures | 15,000 | Capital Y | 50,000 |
| Stocks | 20,000 | | |
| Debtor | 12,000 | | |
| Bank | 8,000 | | |

The assets were revalued on 1 January 2007 to be:

Premises $90,000, Fixtures $11,000, Provision for bad debts $3,000

= 本章完 End of Chapter =

# Chapter 8
## Partnership - Dissolutions

**[01] Introduction**

Legally any change in the constitution of a partnership causes the old partnership to dissolve.

**[02] Reasons**

- The changes in the partnership due to the retirement or death of the partner(s).
- The bankruptcy of the business of the partnership
- The expiration of the partnership agreement.
- The business of the partnership was taken over by a limited company.

**[03] The Procedures of Dissolution**

| Items | Process |
|---|---|
| Assets | • Sold to third party<br>• Taken over by partners in their agreed price |
| Liabilities | • Settle it to someone who are not partners<br>• Sold to third party |
| Loan from partners | Settle it to each partner as distinguished form capital. |
| Current a/c | Transfer the balances to their Capital Accounts. |

**[04] Book-keeping Entries**

| Steps | Descriptions | Book-keeping Entries | |
|---|---|---|---|
| | | Dr | Cr |
| (1) | Transfer provision accounts to relevant assets accounts e.g. Provision for Depreciation | Provision a/c | Assets a/c |
| (2) | Transfer the net book value of assets to Realisation Account, except bank a/c [All assets a/c will be closed] | Realisation a/c | Assets a/c |
| (3) | Assets sold to outsiders | Bank a/c | Realisation a/c |
| (4) | Liabilities which will be taken over by the partners | Liabilities a/c | Capital a/c |
| (5) | Net Assets taken over by partners in agreed price | Capital a/c | Realisation a/c |
| (6) | Payment to the Liabilities | Liabilities a/c | Bank a/c |
| (7) | Discount Received | Creditors | Realisation a/c |

| | | | |
|------|----------------------------------------------|----------------|------|
| (8) | Cost of dissolution | Realisation a/c | Bank |
| (9) | Repayment to Loan | Loan a/c | Bank |
| (10) | Share the profit or loss on Realisation [close the Realisation account] | --- | --- |
| (11) | Settle the credit balances in partners' Capital Accounts to bank account | --- | --- |

*Classwork 1*

Peter and Paul Co had land $100,000, car $12,000 with $3,000 accumulated depreciation, debtors $10,000, creditor $20,000 and bank $5,000. The partnership would be dissolute.

所有 classwork 的工作紙儲存於 CD 內，可以預先列印。

*Classwork 2*

The land sold at $200,000 and other assets were taken over by partners equal.

*Classwork 3*

Receive $9,000 from debtors for settlement the total debts.

*Classwork 4*

Paid $19,000 to creditor for settlement the total debts.

*Classwork 5*

Close all accounts.

(1) Balancing off the "Realisation a/c" and transfer the amounts to "Capital a/c"

(2) Balancing off the "Bank a/c" and transfer the amounts to related "Capital a/c"

(3) All the accounts should be closed and had not any balance carried down.

**[05]  Format of Realisation Account**

### Realisation a/c

| | $ | $ | | $ | $ |
|-----------------------|---|-------|-------------------------|---|-------|
| Book Value of FA | | 1,000 | Capital a/c ( take over) | | |
| | | | ~ A | | 1,000 |
| | | | ~ B | | 2,000 |
| Book Value of CA | | 2,000 | Assets sold | | 3,000 |
| Bank | | 3,000 | Creditors: discount | | 9,000 |
| ~ Cost of dissolution | | | received | | |

| Profit on Realisation | | | | |
|---|---|---|---|---|
| ~A | 3,000 | | | |
| ~B | 3,000 | | | |
| ~C | 3,000 | 9,000 | | |
| | | 15,000 | | 15,000 |

The final balances on the partner's capital accounts should always equal the amount in the bank a/c from which they are to be paid.

**[07] Conclusion**

When a partnership is dissolved, all the assets are sold (or taken over by the partners individually), the liabilities are paid off and whatever remains is paid to the partners, the problems in dissolution accounting are:

(a) calculating the profit or loss on the realisation of the assets;

(b) arriving at the amounts due to or from the partners

Students must be thoroughly familiar with all aspects of the preparation of realisation account, bank account and partners' capital accounts. All these accounts are key accounts of this topics and it's easy to prepare it by studying.

= 本章完 End of Paper =

這課題考試中出現機會很高，要掌握此課，唯一的方法是背熟第 31 頁的步驟。

# Chapter 9
## Limited Company – Issue of Shares

**[01]    Introduction**

| Types of Business 企業類別 | Owners 擁有人 |
|---|---|
| Sole Proprietorship | Sole Trader 東主 |
| Partnership | Partners 合夥人 |
| Non-trading Organisation | Members 會員 |
| Limited Companies | Shareholders 股東 |

At law, a limited company is **an artificial person** and is a separate **legal entity**.
- can own assets
- can bear liabilities in its own name
- can sue its own shareholders or vice versa.

"Limited Liability" means that
- Liability of a shareholder is limited to the nominal value of the shares he holds, whether fully paid or partly paid;
- The liabilities of the company itself are **still unlimited**

**[02]    Different between limited companies and partnerships**

| | Limited Companies | Partnerships |
|---|---|---|
| **No. of owners** | Public: 2 to unlimited<br>Private: 1 to 50 | 2 to 20 (except in certain professional business such as CPA firms) |
| **Amount of capital** | Restricted to the amount of authorized capital | Stated in the partnership agreement |
| **Distribution of Profit** | Dividend | Share to each partners in their agreed ratio |
| **Liabilities** | Limited liability as the contribution amount on their shares held | Unlimited liability of each partner (except limited partners) |
| **Management** | Directors | Partners take part in the operations |
| **Regulating Acts** | Companies Acts | Partnership Act 1890 |
| **Rulers** | Memorandum & Articles of Association | Partnership Agreements |

| Audit | Required | Optional |
|---|---|---|
| **Profit distribution** | Annual General Meeting | As per agreement |

## [03] Private and Public Limited Companies

| | Private Limited Co. | Public Limited Co. |
|---|---|---|
| No of owners | 1 to 50 | At least 2 |
| Share transfer | Agree with shareholders | Traded on HK Stock Exchange |
| Raising fund | ONLY from shareholders | From public |
| No of directors | 1 to 50 | At least 2 |
| Scale | Small | Fever in number and generally much larger in size |
| Governed | Memorandum of Association, Articles of Association | |

## [04] Types of Shares

| Ordinary Shares | Security or voting rights |
|---|---|
| | Have voting rights in general meetings |
| | Income |
| | 1. undistributed profits after senior claims have been made. |
| | 2. Any undistributed profits and all reserves belong to them. |
| | Amount of Capital |
| | 1. all surplus fund after prior claims |
| | 2. They own all reserves |
| Preference Shares | Security or voting rights |
| | Right to vote at general meeting only when it is proposed to change the legal rights of the shares |
| | Income |
| | 1. To paid before an Ordinary Share dividend may be paid. |
| | 2. Interest is not profit tax deductible. |
| | Amount of Capital |
| | A fixed amount per share per year |

Rank of Liquidation

| Liabilities (debentures) | ⇒ | Preferences Shares | ⇒ | Ordinary Shares |
|---|---|---|---|---|

**[05]** **To increase the number of share**

  (a)  Issue to public

| Dr | Cr |
|---|---|
| Bank | Application & Allotment |
| Application & Allotment | Ordinary Share Capital |
| | Share Premium |

*Classwork 1*

On 1 Apr 20X8, Wai Wai Ltd issued 1,000,000 shares to the public at $0.7 (with $0.2 premium) each, payable in full upon application.

Please prepare the ledgers accounts.

所有 classwork 的工作紙儲存於 CD 內，可以預先列印。

  (b)  Issue to existing shareholders - Right Issues
- Give rights certificates to existing shareholders to buy shares at a specified price, and the specified price is normally the market price, so as to reward existing shareholders and to keep existing share-holdings unchanged (only when all shareholders used the right)
- Options of a shareholder:
  (a) ignore it
  (b) accept and buy shares
  (c) sell the right to other people for cash 投機

| Debit (dr) | Credit (cr) |
|---|---|
| Bank | Ordinary Share Capital |
| Share Premium | |

*Classwork 2*

On 20 Apr 20X8, the company received cash of $345,000 from an issue 300,000 shares through a right issue at a premium of $0.15 each.

  (c)  Bonus Issues
- Free shares to existing shareholders, no cash entries
- Why issue? Don't want to distribute cash as dividend
- Proportionate to the existing share-holding, e.g. 1 bonus share for every 5 shares held
- Market value of shares will be reduced from share premium

| Debit (dr) | Credit (cr) |
|---|---|
| Share Premium | Ordinary Share Capital |

*Classwork 3*

On 26 Apr 20X8, the company made a bonus issue of 1 share for every 5 ordinary shares in issue from its sharing premium.

**[06]  Reserves in a company**

| Reserves | = | Net Assets | - | Share Capital |
|---|---|---|---|---|

| | Sources | Distribution Methods |
|---|---|---|
| Revenue Reserves | Net Profit | Dividend |
| Capital Reserves | Issue share ate premium | Bonus Issue |
| | | Discount on share issuing |

**[07]  Provision**

A charge against the profits, to provide for anticipated loss but the amount cannot be accurately determined.

(a)  Provision for diminution in value of assets
- Provision for bad debts
- Provision for depreciation (Level 1)
- Provision for discount allowed

(b)  Provision for liabilities or charge – loss or liability that is likely to be incurred, or is certain to be incurred but uncertain as to amount or date which it will arise
- Provision for long service payment
- Provision for warranty costs
- Provision for compensation

**[08]  Issue Price of raising capital**

(a) Issue at Par

| Issue Price | = | Par Value |
|---|---|---|

*Classwork 4*

| Given Information: Number of shares proposed to issue = 10,000 |
|---|
| Issue Price = $100    Par Value = $100    No. of Application = 12,000 |

(b) Issue at a Premium

| Issue Price | - | Par Value | = | Share Premium |
|---|---|---|---|---|

*Classwork 5*

| Given Information: Number of shares proposed to issue = 10,000 |
|---|
| Issue Price = $120    Par Value = $100    No. of Application = 12,000 |

(c) Issue at a Discount

| Par Value | - | Issue price | = | Share Discount |
|---|---|---|---|---|

*Classwork 6*

| Given Information: Number of shares proposed to issue = 10,000 |
|---|
| Issue Price = $90    Par Value = $100    No. of Application = 12,000 |

Classwork 7

| Issue 1,000 ordinary shares of $0.5 each at a premium of $0.3 on 1 January 20X8. Total application received: 1,200 shares.   Unsuccessful application monies are refunded on 1 Feb 20X8. |
|---|

Bank

| 2008 | | $ | 2008 | | $ |
|---|---|---|---|---|---|
| Jan 1 | Application & Allotment [1] | | | | |

Application & Allotment

| 2008 | | | 2008 | | $ |
|---|---|---|---|---|---|
| Feb 1 | Ordinary Share Capital [2] | | Jan 1 | Bank [1] | |
| | Share Premium [3] | | | | |
| | Bank (refund) [4] | | | | |

## [09]  Case on Application

When a public company asks people to apply for its shares, it does not know how many shares will be applied for.   The company has to say how many shares it wants to issue.

| over-subscription | Sometimes there will be more shares applied for than the company can issue. |
|---|---|
| under-subscription. | At other times, there will be fewer shares applied for than those issued |

The number of application is named as "A". The number of issued is named as "I"

| Relationship between "A" & "I" | Actual number Issued |
|---|---|
| A = I | I |
| A > I | I |
| A < I | A |

Shares Issued at a Premium / Payable in full on application / over-subscription

|  | Dr | Cr |
|---|---|---|
| Received application money | Bank | Applications & Allotments |
| Shares allotted | Applications & Allotments | Ordinary Share Capital<br>Share Premium |
| Refunds to Applicants | Applications & Allotments | Bank |

### Classwork 8

Issue 20,000 ordinary shares of $2 par at $4.   Application together with the necessary money, are received for 30,000 shares. Shares are allotted among the other applicants on a prorata basis, excess application monies are refunded

這課的重點是緊記發行股票的步驟,無論是 Journal 或"T"字帳,都要充份掌握。

= 本章完  End of Chapter =

**[01]    Types of Debentures**

Security or voting rights

1. No voting rights
2. A fixed charge on a specific asset or a floating charge on all assets
3. When liquidate, they entitle to the proceeds of their securities, if proceeds not sufficient, they become unsecured creditors.
4. If surplus, proceeds add to the assets for creditors.

Income

A fixed annual amount whether the company makes a profit or not.

Amount of Capital

A fixed amount per unit of debenture.

**[02]    Issue of Debentures**

a) In general case, the issue of debenture would issue at par.    That means the issuing price is same at the par value.

| Issue at a par | Dr | Cr |
|---|---|---|
| Received application monies | Bank | Debenture Applicants |
| Debentures allotted | Debenture Applicants | Debenture |

*Classwork 1*

On 1 January 20X8, 2,000 $100 10% debentures were issued at par, received in full on application.

所有 classwork 的工作紙儲存於 CD 內，可以預先列印。

b) If the company has the goodwill, the issue of debenture would issue at premium.    That means the issuing price is higher than the par value.

| Issue at a premium | Dr | Cr |
|---|---|---|
| Received application monies | Bank | Debenture Applicants |
| Debentures allotted | Debenture Applicants | Debenture<br>Share Premium |

*Classwork 2*

On 1 January 20X8, 2,000 $100 10% debentures were issued at 110, received in full on application.

c) In some special case such after SARS, the company want easy to issue the debenture, it will issue it at discount. That means the issuing price is lower than the par value.

The difference is so called "Debenture Discount" and it is one kind of financial expense.

| Issue at a discount | Dr | Cr |
|---|---|---|
| Received application monies | Bank | Debenture Applicants |
| Debentures allotted | Debenture Applicants | Debenture |
| | Debenture Discounts | |
| Amortisation of Discounts | Profit & Loss a/c | Debenture Discount |

*Classwork 3*

On 30 Sep 20X8, 3,000 $100 10% debentures were issued at 90, received in full on application. Debenture Discount would be amortized among 5 years on a monthly basis.

**[03]    Issue of Debentures (over subscription)**

If the company has the good reputation, the investors may have interest on its debentures. The quantity demanded of debentures may over than the quantity supplied of it.

a)
| Issue at a par | Dr | Cr |
|---|---|---|
| Received application monies | Bank | Debenture Applicants |
| Debentures allotted | Debenture Applicants | Debenture |
| Refund | Debenture Applicants | Bank |

*Classwork 4*

On 1 January 20X8, 2,000 $100 10% debentures were issued at 100, received 3,000 on application. Applications were received debentures on 5 February.

41

| b) | Issue at a premium | Dr | Cr |
|---|---|---|---|
| | Received application monies | Bank | Debenture Applicants |
| | Debentures allotted | Debenture Applicants | Debenture<br>Share Premium |
| | Refund | Debenture Applicants | Bank |

*Classwork 5*

On 1 January 20X8, 2,000 $100 10% debentures were issued at 110, received in full on application.   Applications were received debentures on 5 February.

| c) | Issue at a discount | Dr | Cr |
|---|---|---|---|
| | Received application monies | Bank | Debenture Applicants |
| | Debentures allotted | Debenture Applicants<br>Debenture Discounts | Debenture |
| | Amortisation of Discounts | Profit & Loss a/c | Debenture Discount |
| | Refund | Debenture Applicants | Bank |

*Classwork 6*

On 30 Sep 20X8, 2,000 $100 12% debentures were issued at 90, received 3,000 applications. Debenture Discount would be amortized among 5 years on a monthly basis.

= 本章完  End of Chapter =

**[01] Introduction 簡介**

For internal purposes, the final accounts of the limited companies are similar as those of the Sole Proprietorship and Partnership with the exception of one expense (e.g. Director's Fee) and the appropriation of net profit (e.g. dividend, transfer to / (from) reserves etc.)

Final Accounts ———————— Internal (similar with "partnership")
———————— External (future for study)

**[02] Special Types of Expenses 特別支出類別**

(a) Debentures Interest / Loan Interest 債券利息 / 貸款利息

| Debenture Interest | = | Debentures | x | Rate (%) |
|---|---|---|---|---|

Trial Balance as at 31 December 20X7

|  | $ | $ |
|---|---|---|
| ………… |  |  |
| 8% Debentures |  | 1,000,000 |
| Debenture Interest | 20,000 |  |

已付債券利息

*Actual Debenture Interest (真正債券利息) in Profit & Loss a/c*
$1,000,000 x 8% = $80,000

*Accrued Debenture Interest (應付債券利息) in Balance Sheet (under the heading of "Current Liabilities")*
Actual Debenture Interest – Paid Debenture Interest
$80,000 - $20,000 = $60,000

*Classwork 1*

Please prepare the Debenture Interest Account.

所有 classwork 的工作紙儲存於 CD 內，可以預先列印。

(b) Director's Fee 董事費 / Director's Emolument 董事酬金

Director's Fee and Director's Emolument are the salaries of the directors. It will be treated as one of administration expenses in the Profit & Loss a/c.

*Classwork 2*

Gross Profit = $100,000, Total Expenses = $78,000
The managing director was entitled to a salary equal to 10% of the net profit (after deduction of his salary). What is the salary of managing director?

(c) Preliminary Expenses / Formation Expenses 開辦費 / 創業費

Preliminary Expenses are the formation expenses of a limited company, including legal fees, professional charges etc.

**[03] Appropriation of Net Profit 盈利的分配**

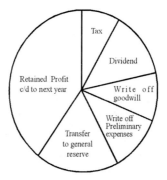

(a) Dividends 股息

Net Profit will be distributed to its shareholders according to the level of net profit (盈利水平) and the dividend policy (派息政策) of the company.

Interim Dividend 中期股息

● paid dividend to the shareholders in the middle of the financial year.
● The amount will be subjected to the performance in the first half of the financial year.

| Dr | Cr |
|---|---|
| Interim Dividend | Bank |
| Appropriation a/c | Interim Dividend |

*Classwork 3*

| Oct 1 | Paid the interim dividend $5,000. |
|---|---|
| Dec 31 | The account clerk prepares the final account. |

Proposed Dividend (BS)/ Final Dividend (A) 擬派股息 / 末期股息

- will be subjected to the shareholders' approvals in the Annual General Meeting (AGM). 週年會員大會

| Dr | Cr |
|---|---|
| Appropriation a/c | Proposed Dividend (Balance Sheet) |

*Classwork 4*

- 5% 10,000 preference shares $1 each & Interim dividend is $300
- What should you to at the end of financial year at 31 Dec 20X7?

Transfer to / from Reserves 撥入或撥出儲備

It meet the future requirements such as Debenture Redemption.

Transfer to Reserve

| Appropriation a/c | ⟹ | Reserves |
|---|---|---|

Transfer from Reserve

| Appropriation a/c | ⟸ | Reserves |
|---|---|---|

Revenue Reserves can be transferred back to Appropriation a/c for Dividend purposes in the future financial period. (Same case of decrease of provision for doubtful debts)

*Classwork 5*

(1) Transfer $10,000 to General Reserve at 31 Dec 20X7.

(2) At the same date, Development Reserve has been decreased by $7,800.

## [04] Appropriation Account

Richard Ltd

Appropriation a/c for the year ended 31 December 20X7.

| | $ | $ | $ |
|---|---|---|---|
| Net Profit before Appropriation | | | 10,000 |
| Less: Appropriation | | | |
| Interim Dividends | | | |
| ~ Ordinary Dividend | 100 | | |

| | | | |
|---|---|---|---|
| ~ Preference Dividend | 200 | 300 | |
| Final Dividends | | | |
| ~ Ordinary Dividend | 200 | | |
| ~ Preference Dividend | 400 | 600 | |
| Reserves | | | |
| Transfer to Redemption Reserve | 1,000 | | |
| Transfer from General Reserve | (500) | 600 | 1,500 |
| Net Profit for the year | | | 8,500 |
| Retained Profit b/d | | | 5,000 |
| Retained Profit c/d | | | 13,500 |

## [05] Classification of Share Capital

(a) Authorised Share Capital (Nominal Capital)

The maximum amount of share that a company is authorised to issue as stated in the Memorandum of Association.

(b) Issued Share Capital

The amount of shares actually issued, it should not exceed authorised share capital

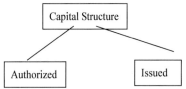

*Classwork 6*

Wai Wai Ltd had an authorised capital of 7,000,000 ordinary share of $0.5 each. On 1 April 20X8, it has already issued 3,000,000 ordinary shares and decided to issue all the remaining un-issued shares at $1 per share.

## [06] Rights of shareholders

- To vote in the meeting to appoint directors
- To attend the Annual General Meeting
- To receive dividend (the amount given to shareholders as their share of profit)

## [07] Other means of funding

| Debentures | Reserves |
|---|---|
| • Long-term Loans | • Issue share of premium |
| • Fixed rate of interest payable | |
| • Fixed date of redemption | |

## [08] Balance Sheet

The Balance Sheet of the limited companies are similar 近似 as those of the Sole Trader 獨資經營 and Partnership 合夥經營 with the exception 除了 of the "Share Capital"股本 and "Reserves" 儲備 sections

Richard Happy Ltd

Balance Sheet (Extract) as at 31 December 2007

| | $ | $ |
|---|---|---|
| Financed by: | | |
| Authorised Share Capital 註冊股本 | | |
| 2,000,000 Ordinary Shares of $2 each | | XXX |
| 1,000,000 Preference Share of $0.5 each | | XXX |
| | | XXX |
| | | |
| Issued Share Capital 發行股本 | | |
| 1,000,000 Ordinary Shares of $2 each, fully paid | | XXX |
| 500,000 Preference Share of $0.5 each | | XXX |
| | | XXX |
| | | |
| Reserves 儲備 | | |
| Share Premium 溢價 | XXX | |
| General Reserve 一般儲備 | XXX | |
| Retained Profit (c/d) 留存利潤 | XXX | |
| Proposed Dividend 擬派股息 | XXX | XXX |
| | | XXX |

*Classwork 7*

| |
|---|
| (a) Net Profit before Appropriation $100,000 |
| (b) Transfers to Debenture Redemption Reserve $50,000 |
| (c) Transfer from General Reserve $20,000 |
| (d) Interim Dividend (i) Ordinary $5,000 (ii) Preference $6,000 |
| (e) Final Dividend (i) Ordinary $10,000 (ii) Preference $12,000 |
| (f) Retained Profit b/d $50,000 |

= 本章完 End of Chapter =

## [01] Manufacturing Account 製造成本帳

Some of the firms manufacture the goods by themselves instead of purchases the goods from the suppliers.   In general, it can minimize the **cost of good sold** in the Trading a/c in the long run.   Therefore, **Manufacturing a/c** is prepared before the Trading a/c.

Comparison:

| Account | Functions |
|---|---|
| Trial Balance | For checking contain balances of the Asset & Liability a/c |
| Trading a/c | To find out Gross Profit or Gross Loss |
| Profit & Loss a/c | To find out Net Profit or Net Loss |
| Appropriation a/c | To find out Retained Profit c/d |
| Manufacturing a/c | To find out Production cost |

Production Cost of Goods Completed 完成品之生產成本

Manufacturing a/c

| | $ | | $ |
|---|---|---|---|
| Costs of goods manufactured | X | Value of goods manufactured | A |

Trading a/c

| | $ | | $ |
|---|---|---|---|
| Value of goods manufactured | A | | |

| Prime Cost | | Factory Overheads | | Work In Progress (start) | | Work In Progress (end) |
|---|---|---|---|---|---|---|
| M1 | *Add* | M2 | *Add* | M3 | *Less* | M3 |

承接上一時期
未完成的貨品

留給下一时
期來完成

| | | $ |
|---|---|---|
| | What is the production cost of goods completed? | |
| M1 | Prime Cost 主要成本 | 100 |
| | Factory Overheads 間接費用 | 200 |
| | Work in Progress – 1.1.20X7 半製成品 | 300 |
| | Work in Progress – 31.12.20X7 | 200 |
| | Rent 租金 | 300 |
| | Sales 銷貨 | 1,000 |
| T | Return Inwards | 200 |

所有 classwork 的工作紙儲存於 CD 內，可以預先列印。

## [02] Prime Cost 主要成本 (M1)

Prime Cost is the total direct cost in the progress of manufacturing that can be directly traced to the items manufactured. [Prime Cost 主要成本 → 3D]

(A) Direct materials are the costs of raw materials used in the production of goods such as rubber for sport shoes

| Opening Materials | + | Purchases | - | Return Outwards | + | Carriage Inwards | - | Closing Materials |

(B) Direct labour is the cost of manpower involved in the production line. Wages paid to people not directly engaged in production but assisting in production are called " indirect labour costs"

(C) Direct expenses are the costs of expenses that can allocated to each goods such as trademark, patent, etc.

Which is the Prime Cost?                                          $

| | | |
|---|---|---|
| M1 | Opening Materials 期初原料 | 100 |
| | Purchase – finished goods 購買完成品 | 200 |
| | Purchase – raw materials 購買原料 | 300 |
| | Return Outwards – finished goods 完成品退貨 | 100 |
| | Return Outwards – raw materials 原料退貨 | 50 |
| | Carriage Inwards – raw materials 原料購買運費 | 100 |
| T | Closing Stock – finished 期末存貨 | 50 |
| | Closing Stock – raw materials 期末原料 | 100 |
| | Direct Manufacturing Wages 直接工資 | 400 |
| | Factory Manager Wages 工廠經理工資 | 500 |
| | Office Manager Salary 辦公室經理薪金 | 1,000 |
| | Bonus of CEO 行政總裁獎金 | 10,000 |
| | Royalties 專利權 | 600 |
| PL | Advertising 廣告 | 700 |

所有屬於原料都是 M1

## [03] Factory Overheads 間接製造成本 (M2)

Factory Overheads are all total indirect expenses in the process of manufacturing that cannot be traced to the items manufactured.

- Indirect materials – materials for repairs of machinery
- Indirect labour – wages to manager, cleaners
- Indirect expenses – provision for depreciation - machinery

Wages paid to people not directly engaged in production but assisting in production are called "indirect labour costs" such as wages paid to work supervisors storemen and time keepers

In the public examination and actual practice, some of the expenses items will be apportioned between the factory and office of the company. The former will go to the Manufacturing a/c while the latter will go to the Profit & Loss a/c.

*Classwork 3*

| | |
|---|---|
| Rent & Rates 租金及差餉 | $400,000 |
| Insurance 保險 | $15,000 |
| Heating & Lighting 電費 | $36,000 |
| Telephone 電話費 | $6,000 |

Provision for Depreciation 折舊準備

| | |
|---|---|
| ~ office equipment 辦公室設備 | $1,000 |
| ~ machinery 機器 | $2,000 |
| ~ motor vehicle 小型汽車 | $4,000 |

Additional Information: 附加資料

- Insurance included $3,000 for the last years.
- Accrual telephone Expenses $500 本年度應付電話費開支$500
- 東主所住的租金 $10,000 Sole Trader's private rent $10,000
- Half time of motor vehicles was used in office.
- Expenses should be apportioned as follows:

| | M2 Factory 工廠 | PL Office 辦公室 | Total $ |
|---|---|---|---|
| Rent & Rates | 3/4 | 1/4 | |
| Heating & Lighting | 3/4 | 1/4 | |
| Insurance | 5/6 | 1/6 | |
| Telephone | 1/2 | 1/2 | |
| Provision for Depr. | --- | --- | |

## [04] Work in Progress 半製成品 (WIP) (M3)

Work in Progress is the semi-finished goods in the manufacturing process that cannot be sold. 半製成品是生產過程中未完成的產品。

The opening WIP is to be continuously worked on absorption further costs until it becomes completed goods. In fact, the opening WIP is "one of the resources used in the production of completed goods in the current period."

$$\boxed{M1} + \boxed{M2} + \boxed{\text{Opening WIP}} - \boxed{\text{Closing WIP}} = \boxed{\text{Production Cost}}$$

## [05] Format of Manufacturing a/c 製造成本帳的格式

A firm engaged in manufacturing as well as trading activities will have a Trial Balance containing the balances of all accounts pertinent to the Manufacturing a/c and the Trading & Profit & Loss a/c. Needless to say, the trial balance will also contain balances of the Asset and Liability accounts.

> M1 – Prime Cost / M2 – Overheads / M3 – WIP

Richard Ltd
Manufacturing a/c for the year ended 31 December 20X7

| | $ | $ | $ |
|---|---|---|---|
| Direct Materials 直接物資 | | | |
| Opening Stock 期初存貨 | | 1,000 | |
| Purchase 購貨 | 500 | | |
| Return Outwards 購貨退出 | (100) | | |
| Carriage Inwards 購貨運費 | 200 | 600 | |
| Closing Stock 期末存貨 | | (400) | 1,200 |
| Direct Labour 直接人工 | | | |
| Manufacturing Wages 製造工資 | | | 2,000 |
| Direct Expenses 直接費用 | | | |
| Copyrights 版權 | | | 3,000 |
| **Prime Cost 主要成本** | | | 6,200 |
| **Factory Overheads 間接費用** | | | |
| Indirect Wages , Rent & Rates, Electricity | | 3,000 | |
| Provision for Depreciation – Machinery 折舊準備 | | 2,000 | 5,000 |
| | | | 11,200 |
| Work In Progress 1.1.2007 半製成品 | | | 500 |
| | | | 11,700 |
| Work In Progress 31.12.2007 半製成品 | | | (1,000) |
| Production Cost of Goods completed 生產成本 (轉移往購銷帳) | | | 10,700 |

**[06]    Cost of Goods completed transfer to Trading a/c**

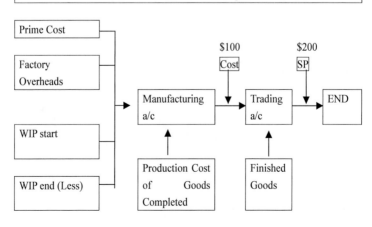

52

|  | $ | $ |
|---|---|---|
| Finished Goods Transfer to Trading a/c<br>生產成本移至購銷帳 |  | 2,500 |
| Sales 銷貨 |  | 20,000 |
| Less: Cost of Good Sold 減：銷貨成本 |  |  |
|     Opening Stock 期初存貨 | 1,000 |  |
|     Production Cost 生產成本 | 2,500 |  |
|     Purchase 購貨 ($20/unit) | 4,000 |  |
|     Closing Stock 期末存貨 | (2,000) | 5,500 |
| Gross Profit 毛利 |  | 14,500 |
| Add: Other Income 加：其他收入 |  |  |
|     Interest 利息 |  | 1,000 |
|  |  | 15,500 |
| Less: Administration Expenses |  |  |
|     電費 Electricity | 100 |  |
|     Provision for depreciation - Vehicle | 200 |  |
|     Insurance 保險費 | 300 | 600 |
|     Selling & Distribution Expenses |  |  |
|     Sales Commission |  |  |
| Net Profit 純利 |  | 14,900 |

## [10] Treatment of closing stock 期末存貨處理

In a manufacturing business, there are three types of closing stock as follows:

| Types | Final Accounts | Treatment of Closing Stock in Balance Sheet | |
|---|---|---|---|
| Raw Materials | Manufacturing a/c | Transfer at Production Cost of Goods Completed | |
| WIP | Manufacturing a/c | Current Assets | $ |
| Finished Good | Trading a/c | Stock | |
|  |  | ~ raw materials | X |
|  |  | ~ WIP | X |
|  |  | ~ finished goods | X |
|  |  |  | X |

**[11] Expenses charged to profit and loss a/c**

a) Administration expenses

These expenses do not be affected by the sales volume.

- The salaries of directors, management and office staff;
- Rent and Rates;
- Insurance;
- Telephone and postage;
- Printing and stationery;
- Heating and lighting.

b) Selling and distribution expenses

There are expenses associated with the process of selling and delivering goods to customers.

- The salaries of a sales director and sales management.
- The salaries and commissions of salesmen.
- The traveling and entertainment expenses of salesmen.
- Marketing costs(e.g. advertising and sales promotion expenses).
- The costs of running and maintaining delivery vans.
- Discounts allowed to customers for early payment of their debt.
- Bad debts written off.
- Provision for Depreciation – Office Equipment

c) Finance expenses

For example: Interest on a loan; Bank overdraft interest.

= 本章完 End of Chapter =

## [01] Introduction

**Suspense account** is an account showing a balance equal to the difference in the trial balance.

## [02] Suspense account and the balance sheet

If the errors are not found before the final accounts are prepared, the suspense account balance will be included in the balance sheet. When the balance is a credit balance, it should be included on liabilities side of the balance sheet.

## [03] Correction of errors – one error only

**Example 1:** Assume that the error of $40 as shown in was found in the following year on 31 March 20X8. The error was that the sales account was added up incorrectly and shown as $40 too little.

Suspense a/c

| 20X8 | | $ | 20X8 | | $ |
|---|---|---|---|---|---|
| Mar 31 | Sales | 40 | Jan 1 | Balance b/d (bal) | 40 |
| | | 40 | | | 40 |

Sales

| 20X8 | | $ | 20X8 | | $ |
|---|---|---|---|---|---|
| Mar 31 | | | Jan 1 | Suspense | 40 |

**Example 2:** In another business, the trial balance on 31 Mar 20X8 had a difference of $168. It was a shortage on the debit side of the trial balance. A suspense account was opened.

Suspense a/c

| 20X8 | | $ | 20X8 | | $ |
|---|---|---|---|---|---|
| Mar 31 | Bal b/d (bal fig) | 168 | Jan 1 | Trial Balance | 168 |
| | | 168 | | | 168 |

**Example 3:** On 31 Mar 20X8 the error was found.   We had made a payment of $168 to K. Lee to close his account.   It was correctly entered in the cash book, but it was not posted to K. Lee's account.

Suspense a/c

| 20X8 | | $ | 20X8 | | $ |
|---|---|---|---|---|---|
| Mar 31 | Bal b/d (bal fig) | 168 | Mar 31 | K. Lee | 168 |
| | | 168 | | | 168 |

K. Lee

| 20X8 | | $ | 20X8 | | $ |
|---|---|---|---|---|---|
| Mar 31 | Suspense (correction) | 168 | Mar 31 | Bal b/d | 168 |
| | | 168 | | | 168 |

## [04]   Correction of error – more than one error

**Example 4:** The trial balance as at 31 Mar 20X7 showed a difference of $77, being a shortage on the debit side.   A suspense account was opened.   On 28 February 20X8 all the errors from the previous year were found.

(i)     A cheque of $150 paid to Richard's had been correctly entered in the cash book, but had not been entered in Richard's account.

(ii)    The purchases account had been undercast by $20.

(iii)   A cheque of $93 received from K. San had been correctly in the cash book, but had not been entered in San's account.

Richard

| 20X8 | | $ | 20X8 | | $ |
|---|---|---|---|---|---|
| Feb 28 | Suspense (correction) | 150 | Feb | Bal b/d | 150 |
| | | 150 | | | 150 |

Suspense a/c

| 20X8 | | $ | 20X8 | | $ |
|---|---|---|---|---|---|
| Feb 28 | Bal b/d | 77 | Feb 28 | Richard | 150 |
| | K San | 93 | | Purchases | 20 |
| | | 170 | | | 170 |

### Purchases

| 20X8 | | $ | 20X8 | | $ |
|---|---|---|---|---|---|
| Feb 28 | Suspense | 20 | Feb 28 | Trading | 20 |
| | | 20 | | | 20 |

### K. San

| 20X8 | | $ | 20X8 | | $ |
|---|---|---|---|---|---|
| Feb 28 | Bal b/d | 93 | Feb 28 | Suspense | 93 |
| | | 93 | | | 93 |

## [06] Profit adjustment

### Example 5:

Net Profit $10,600

1. An amount paid for advertising $128 had been correctly entered in the Cash Book, but entered as $281 in the Advertising Account.
2. The balance of $347, on a customer's account, had been written off the Debtors Account, but the double entry had not been completed.
3. Discount allowed $223 had been correctly entered in Cash Book and in customer's account, but had been credited to the Discount Received Account.

Calculate the revised profit for the year ended 31 March 2006

| | Dr | Cr | Increase | Decrease |
|---|---|---|---|---|
| (1) | | | | |
| (2) | | | | |
| (3) | | | | |

= End of Paper 本章完 =

| Chapter 14 |
|:---:|
| Control a/c |

## [01] Introduction

Control Account is an account, which checks the arithmetical accuracy of a ledger.

Transactions

| Credit Sales | Credit Purchase | Return Inwards | Return Outwards | Cash Purchase, Cash Sales, Paid Creditors, Received from Debtors, Cash drawings, Cash payment, Cash receipts | Bad Debt, Depreciation, Buy FA on credit, Corrections of errors … | Sundry daily exp. |
|---|---|---|---|---|---|---|
| Sales Journal | Purchases Journal | Return Inwards Journal | Return Outwards Journal | Cash Book | The Journal | Petty Cash Book |

"T" account in Sales Ledger/Purchase Ledger/General Ledger

Control accounts are sometimes known as "total accounts". It is an easy way to find out the balance of creditors and debtors. Thus, a control account for a Sale Ledger could be known either as a "sale ledger control account" or as a "total debtor account".

58

| Creditors' Account | ——————→ | Purchase Ledger Control Account (PLC) |
|---|---|---|

## [02] Functions and advantages of control accounts

- To ensure the arithmetical accuracy of the entries
- For internal checking (department heads)
- To find out total debtors' and total creditors' balance easily without reference to sales ledger and purchase ledger accounts.
- To locate the difference with the aid of the control accounts when two sides of a trial balance do not agree
- Control accounts are kept in the general ledger
- Subsidiary ledger consisting of individual debtors/creditors accounts (individual ledger cards) is kept to support the control account.
- Control accounts can provide debtors/creditors balances quickly to facilitate the preparation of trial balance and balance sheet
- It shows a separation of duties in keeping the general ledger and subsidiary ledgers

## [03] Information for control accounts

| Sales Ledger Control | Source |
|---|---|
| 1. Opening Debtors | The sum of balance b/d on all debtors. |
| 2. Credit Sales | Total from Sale Journal / Sales Day Book. |
| 3. Return Inwards | Total of Return Inwards Journal. |
| 4. Cheque received | Bank Column of Cash Book on received side. |
| 5. Cash received | Cash Column of Cash Book on received side. |
| 6. Discount Allowed | Total Discount Allowed Column of Cash Book on received side. |
| 7. Closing Debtors | The sum of balance c/d on all debtors |

Sales Ledger Control a/c

| | $ | | $ |
|---|---|---|---|
| Bal b/d | | Return Inwards | |
| Sales | | Bank | |
| | | Cash | |
| | | Bad Debt | |
| | | Discount Allowed | |
| | | Bal c/d | |

| Purchase Ledger Control | Source |
|---|---|
| 1. Opening Creditors | The sum of balance b/d on all creditors. |
| 2. Credit Purchase | Total from Purchase Journal / Purchase Day Book. |
| 3. Return Outwards | Total of Return Outwards Journal. |
| 4. Cheque Paid | Bank Column of Cash Book on payments side. |
| 5. Cash Paid | Cash Column of Cash Book on payment side. |
| 6.Discount Received | Total Discount Received Column of Cash Book on payment side |
| 7. Closing Creditors | The sum of balance c/d on all creditors. |

Purchase Ledger Control a/c

| | $ | | $ |
|---|---|---|---|
| Bank | | Bal b/d | |
| Cash | | Purchase | |
| Discount Received | | | |
| Bal c/d | | | |

*Classwork 1*

| | |
|---|---|
| Debit Balance on 1 Jan 20X8 | 20,000 |
| Total Credit Sales for the month | 100,000 |
| Cheques received from customers in the month | 70,000 |
| Cash received from customers in the month | 12,000 |
| Returns inwards from customers during the month | 2,000 |
| Debit balance on 31 Jan as extracted from the sales ledger | 33,000 |
| Total Cash Sales | 50,000 |
| Dishonor cheque | 1,000 |
| Bad Debt Recover | 3,000 |
| Bad Debt | 4,000 |

所有 classwork 的工作紙儲存於 CD 內，可以預先列印。

> If the total of a control account are not equal and the entries made to it were correct, this shows that there is an error somewhere in the ledger.

*Classwork 2*

| | |
|---|---|
| Credit Balance on 1 Jan 20X8 | 10,000 |
| Total Credit Purchase for the month | 100,000 |

| | |
|---|---|
| Cheques paid to suppliers in the month | 70,000 |
| Returns outwards to suppliers during the month | 2,000 |
| Credit balance on 31 Jan as extracted from the purchase ledger | ??? |
| Cash Purchase | 2,000 |
| Discount Received | 1,000 |

## [04]  Set-Off

A contra account whereby the same firm is both supplier and a customer and inter-indebtedness is set-off, will also need entering in the control accounts.

In General Ledger

Dr: Purchase Ledger Control          Cr: Sales Ledger Control

*Classwork 3*

Boz owed $10 to Apple.   Apple owed $7 to Boz. Accounts settled by transfer from one ledger to other $7. In the Boz's book

*Classwork 4*

What is the entry of Apple's book?

## [05]  Errors in sale ledger control

1. prime entry records are incorrect
2. adjustments not posted to sale ledger control account
3. the sales ledger control account may have been incorrectly balanced at the ended of the accounting period

## [06]  Errors in sale ledger

1. entries have been missed, or incorrectly made on individual customer accounts
2. individual debtor accounts might have been incorrectly balanced at the end of the accounting period
3. one or more debtor account balances might have been omitted from the listing of the sales ledger

| Errors | Corrections | |
|---|---|---|
| | Dr | Cr |
| (a)  Credit Sales Invoice omitted *(Increase the debtors ledger listing)* | SLC | Sales |

| | | |
|---|---|---|
| (b) Sales omitted in journal | SLC | Sales |
| *(Increase the debtors ledger listing)* | | |
| (c) Sales of $12 incorrectly entered n sales journal as $10 | SLC | Sales |
| *(Increase the debtors ledger listing)* | | |
| (d) Sales journal undercast | SLC | Sales |
| *(The debtos ledger listing remain constant, because only the figure is wrong. This does not affect the debtor accounts, the figures for which are entered individually.)* | | |
| (e) Sales of $12 incorrectly entered as $10 in sales accounts | Suspense | Sales |
| (f) Sales understated in SLC | SLC | Suspense |
| (g) Sales of $12 incorrectly debited to debtor account as $10. | No Entry | |
| (h) Balance of sales has been incorrectly put to the wrong side of the trial balance | Suspense | --- |
| (i) Balance of debtors has been incorrectly put to the wrong side of the trial balance | No entry | |
| (j) Balance of SLC a/c has been incorrectly put to the wrong side of the trial balance | --- | Suspense |

= 本章完  End of Chapter =

這課題是第二級最艱深的一課

# Chapter 15
## Accounting Ratios & the Interpretation of Accounts

**[01] Introduction**

The following bases may be used to compare the accounting ratios of current financial period.

1. The accounting ratios of last financial period
2. The budgeted accounting ratios of current financial period
3. The accounting ratios of other companies in the same industry

Can be used to ascertain the performance of a business in a particular period, or to provide a basis for comparing the results of the current year with those of previous or for comparing the results of different companies in the same industry.

**[02] Use of Ratio Analysis**

Ratio analysis is a process of evaluating the relationship between the component parts of financial statements to obtain a better understanding of a firm's performance and financial position.

By ratio analysis, the performance of the firm can be evaluated:

| | | |
|---|---|---|
| 1. | profitability, | Level 2 |
| 2. | liquidity | Level 2 |
| 3. | management efficiency | Level 2 & Level 3 |
| 4. | investment return | Level 3 |
| 5. | financial stability | Level 3 |

Parties interested in analyzing financial statements

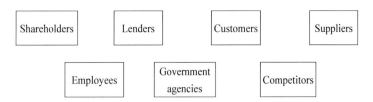

*Classwork 1*

Why they want to know the ratios?

所有 classwork 的工作紙儲存於 CD 內，可以預先列印。

*Classwork 2*

If commission is 5% of profit after commission is deducted, and profit before commission is $16,800, then commission is ?

## [02] Markup & Margin

$$\text{Mark up} = \frac{1}{3}$$

$$\text{Margin} = \frac{1}{3+1} \Rightarrow \frac{1}{4}$$

- With same numerator of markup 分子相同
- Denominator = mark up's denominator + numerator 分母會加上分子

*Classwork 3*

If mark up is 25%, what is margin?

*Classwork 4*

If margin = 20%, what is mark up?

## [03] Calculating Missing Figures

In examination, the main use of Ratios is……… calculating missing figure.

*Classwork 5*

All goods in a firm have the same rate of markup and ignore wastage and theft of goods.  What is the amount of sales? [Hints: Open the trading account]

Figures for 20X7

| | |
|---|---|
| Stock 1.1.20X7 | $400 |
| Stock 31.12.20X7 | $600 |
| Purchase | $5,200 |
| A uniform mark-up rate 20% | |

## [04] Calculation of Accounting Ratios – Profitability

To measure the performance of management and to identify whether the company is a worthwhile investment opportunity relative to its competitors

(i)   Gross Profit Ratio / Gross Profit Margin

$$\text{Margin} = \frac{\text{Gross Profit}}{\text{Net Sales}} \times 100\%$$

- It shows the amount of GP for every $100 of sales

- Sales increased -→ GP % fallen
- It indicates the efficiency of cost control & pricing policy of company
- A higher gross profit ratio may indicate the goods are being sold at higher
- Sometimes the company may lower its gross profit margin in order to stimulate sales (law of demand)
- A low gross profit ratio may not result in a small figure of gross profit unless a small volume of sales accompanies it.
- It can defect the errors and fraud affecting the items in the trading account
- The staff of Inland Revenue to defect the suppression of sales uses it.

*Classwork 6*

Why the gross profit ratio would be decrease?

(ii) Net Profit Ratio

$$\text{Net Profit Ratio} = \frac{\text{Net Profit}}{\text{Net Sales}} \times 100\%$$

- It indicates the efficiency of expenses control
- Sometimes a high gross profit ratio with a low net profit ratio may indicate too much money is spent on the operating expenses. Investigation should be made on the operating expenses to find out which expense occupied a high proportion.

*Classwork 7*

High margin, low Net Profit Ratio. What does it mean?

(iii) Return on Capital Employed

$$\text{ROCE} = \frac{\text{EBIT}}{\text{Total Assets} - \text{CL}} \times 100\%$$

- The most important ratio of all
- It indicates the efficiency of management in generating profits from the capital employed
- The higher the ratio, the more profit the company gained
- It enables the profitability of the company comparable with the other investment opportunities.

| Classwork 8 | Company A | Company B |
|---|---|---|
| Net Profit for the year | 50 | 65 |
| Share Capital | 100 | 150 |
| General Reserve | 60 | 50 |
| Retained Profit | 40 | 75 |
| Debentures | 50 | 50 |
| Calculate the return on total capital employed? Which company is more profitable? | | |

- In exam, use the method stated by the examiner

## [05] Calculation of Accounting Ratios – Liquidity

A company's liquidity is its ability to pay its debt by having enough cash or liquid assets easily available

(i) Current Ratio / Working Capital Ratio

$$\text{Current Ratio} = \frac{CA}{CL} : 1$$

- A high current ratio may indicate the company has sufficient current assets to meet its current liabilities
- A low current ratio may indicate the company may face the liquidity problems (not be able to meet its current liabilities)

(ii) Quick ratio (Acid test ratio)

$$\text{Quick Ratio} = \frac{CA - \text{stock}}{CL} : 1$$

- to indicate the ability of a firm to meet it's pay out of its Quick Assets[1].
- A high quick ratio may indicate the company has sufficient liquid assets to meet its current liabilities

Classwork 9

|  |  | ABC Ltd | DEF Ltd |
|---|---|---|---|
| Current Assets: | Stock | 300 | 300 |
|  | Debtors | 200 | 200 |
|  | Bank | 100 | 100 |

---
[1] quick assets mean the assets which can be converted into cash very quickly

| | | |
|---|---|---|
| Current Liabilities: Creditors | 400 | 200 |
| Working Capital Ratio & Quick Ratio | | |

- If the quick ratio is too high, the company may be holding too much idle liquid assets such as debtors or cash
- A low quick ratio may indicate the company may face the liquidity problems
- The difference between current ratio and quick ratio may indicate the proportion of stock holding. The greater the difference may indicate the company is holding too much stock.

*Classwork 10*

What is the solution to improve the liquidity position?

## [06] Calculation of Accounting Ratios –Activity & Efficiency

Efficiency ratios are used to assess the efficiency of financial management of a company such as stock control, credit control and asset management.

(i) Stock Turnover

$$\text{Stock Turnover Rate} = \frac{\text{Cost of Good Sold}}{\text{Average Stock}} = \text{X times}$$

$$\text{Stock Turnover Rate} = \frac{\text{Average Stock}}{\text{Cost of Good Sold}} \times 365 \text{ days} = \text{X days}$$

$$\text{Stock Turnover Rate} = \frac{\text{Average Stock}}{\text{Cost of Good Sold}} \times 52 \text{ weeks} = \text{X weeks}$$

$$\text{Stock Turnover Rate} = \frac{\text{Average Stock}}{\text{Cost of Good Sold}} \times 12 \text{ months} = \text{X months}$$

- It shows how well the firm is managing to do these things.
- Any increase in stock or slowdown in sales will show a lower ratio.
- If the opening stock is not given then the closing stock is assumed to be the average stock.

*Classwork 11*

What does it mean the low stock turnover?

*Classwork 12*

How to improve the stock turnover?

*Classwork 13*

| | |
|---|---|
| 1. Stock Turnover = 5 times<br>Opening Stock = $600<br>Closing Stock = $900<br>Purchase = | 2. Stock Turnover = 12imes<br>Opening Stock = $2,000<br>Closing Stock =<br>Purchase = $16,000 |

A firm should try to do the following

(i) Keep its stock to as low a figure as possible without losing profitability.

(ii) Sell its goods as quickly as possible.

(ii) Turnover of Debtors / Credit Period Allowed to Debtors

$$\text{Turnover of Debtors} = \frac{\text{Trade Debtors}}{\text{Credit Sales}} \times 365 \text{ days} = \text{X days}$$

$$\text{Turnover of Debtors} = \frac{\text{Trade Debtors}}{\text{Credit Sales}} \times 52 \text{ weeks} = \text{X weeks}$$

$$\text{Turnover of Debtors} = \frac{\text{Trade Debtors}}{\text{Credit Sales}} \times 12 \text{ months} = \text{X months}$$

- This ratio assesses how long it takes for our debtors to pay us
- The shorter the debtors collection period the better
- Sometimes, the company may offer generous credit terms to its customers in order to stimulate sales
- The higher the ratio, the worse we are at getting our debtors to pay on time the lower the ratio, the better we are at managing our debtors.

*Classwork 14*

Why the firm wants debtors pay faster?

(iii) Turnover of Creditors / Credit Period Received from Creditors

$$\text{Turnover of Creditors} = \frac{\text{Trade Creditors}}{\text{Credit Purchases}} \times 365 \text{ days} = \text{X days}$$

$$\text{Turnover of Creditors} = \frac{\text{Trade Creditors}}{\text{Credit Purchases}} \times 52 \text{ weeks} = \text{X weeks}$$

$$\text{Turnover of Creditors} = \frac{\text{Trade Creditors}}{\text{Credit Purchases}} \times 12 \text{ months} = \text{X months}$$

- This ratio shows how long it takes a firm on average to pay its suppliers
- The lower creditors turnover the better

*Classwork 15*

Why for a firm to take longer time to pay its suppliers?

(iv) Assets Turnover Ratio

$$\text{Assets Turnover Ratio} = \frac{\text{Net Sales}}{\text{Total Assets}} = \text{X times}$$

- It indicates the efficiency of utilization of total assets in generating revenue
- A low capital (assets) turnover may indicate:
    1. an inefficient management and utilization of assets
    2. an excessive investment in assets

*Classwork 16*

How to improve the capital (assets) turnover?

(v) Debtors : Creditors Ratio

The ratio compares how much credit is allowed by the business to its customers (debtors), and how much credit has been received by the business from its suppliers (creditors).

**[07] Limitations of Ratio Analysis**
- Based on historical cost - it may not be relevant for predicting future performance.
- Different firms may adopt different accounting policies which make comparisons difficult such as depreciation on SL or RB
- The uniqueness of a particular year or a particular firm has to be considered before a meaningful comparison can be made.

- Changes in the value of money and the price levels may weaken the validity of comparisons of ratios computed for different years such as the appreciation of RMB since 2006 and inflation in 2008
- Since the financial statements are prepared at a date, they do not disclose the short-term fluctuations during the accounting period.

**[08] Overtrading**
- It occurs when there is insufficient control over working capital resulting in insufficient liquid funds to meet the demands of creditors
- It is generally the result of too fast sales growth in relation to level of debtors, creditors and stock
- If it occur: raise long-term finance, or cut-back on the expansion

**[09] Sample of interpretation of Accounting Ratios**

1.

| Profitability | Firm A | Firm B | Remarks |
|---|---|---|---|
| Gross Profit Margin | | | |
| Net Profit Margin | | | |
| Return on Capital Employed | | | |

Analysis:

1. By comparing their ROCE, Firm A is (more profitable), (50)% compared with (29)% of Firm B. → Firm A is (more efficient) in utilizing the capital to generate revenue.
2. Firm A has (higher profit margin) of the sales, (42)% compared with (24)%. → use high (profit margin policies)
3. Firm A's NP% (drop more) firm GP% than Firm B's, from (42)% to (15)%, compared with (24)% to (10)% of Firm B. → Firm B is (more efficient) / has better control of.

2.

| Liquidity | Firm A | Firm B | Remarks |
|---|---|---|---|
| Current Ratio | | | |
| Quick Ratio | | | |

70

Analysis:

1.  Current Ratio: both companies are in acceptable range, (2) to 1 and (3.1) to 2. → able to repay (debts) when it is done and have sufficient (working capital) for further investment and operation. → Firm B has (more cash) pay back its debts as its ratio is higher (1.3) to 1 compared with (0.75) to 1 of Firm B.
2.  Quick Ratio: Firm A has higher ability to pay (urgent payment), Firm B <1 → (poor liquidity position)
3.  Compare 2 ratios: Firm B has higher current ratio but lower quick ratio → (build up) in stock level / (accumulation) of stock → (trading slows down)

| 3. | Efficiency | Firm A | Firm B | Remarks |
|---|---|---|---|---|
| | Stock Turnover Ratio | | | |
| | Debtors Collection Period | | | |
| | Creditors Repayment Period | | | |

Analysis:

1.  *Firm A has (better stock level control), it rate is (3.4) times than (2.9) times of Firm B.*
2.  *Firm B has higher risk of (stock accumulation) → refer to quick ratio*
3.  *We should also consider:*
*   *(the nature of business)*
*   *(industry average)*
4.  *Firm B has better_____, it takes (1.8) months to collect its debt while Firm A takes (2.2) months.*
5.  *Firm A can settle its debt in _____. (5.1) months compared with _____ months → better _____*
6.  *If the period is too long → risk of _____ and _____*
7.  *Compare two ratios: Firm A can have more _____ in the company*

**[09] Examination Techniques**

1. When calculating the ratios, you should:
- State the formula
- Substitute the suitable figure into the formula
- Computer the result
- Add the correct unit

2. When interrupting the ratios, you should:
- State out which firm/year is better in profitability or liquidity
- Support your answer with ratios
- Compare the ratios and give simple explanation

= End of Chapter = 本章完

# Chapter 16
## Incomplete Records & Single Entry

**[01] Introduction**

The owner may only maintain the information as follows by single entry system:

| | |
|---|---|
| Cash / bank receipts and payments | Fixed assets |

| | |
|---|---|
| Debtors | Creditors |

Reasons for recording in single entry system
- Limited resources for some small business
- Firms would enter details of a transaction once only
- Many of the businesses fail to record every transaction resulting in incomplete records.

Disadvantages:
1. The records of business transactions are incomplete
2. The final accounts cannot be produced quickly and accurately because of the absence of nominal accounts
3. The final results are not reliable because the operating results of the company are estimated based on the increase or decrease of capital in the period.
4. Errors and frauds may occur and are difficult to detect

**[02] Calculation of Net Profit / (Loss) from Statement of Affairs**

a) Statement of Affairs is same as balance sheet

| Statement of Affairs (incomplete records) | = | Balance Sheet (double entry system) |
|---|---|---|

Richard Ltd
Statement of Affairs as at 31 December 2007

| Assets | $ | $ |
|---|---|---|
| Fixed Assets | 4,000 | |
| Currents Assets | 5,600 | 9,600 |
| Less: Liabilities | | |
| Current Liabilities | 1,500 | |
| Long Term Liabilities | 3,600 | 5,100 |
| Capital as at 31 Dec 2007 | | 4,500 |

b) Find Net Profit by the movement of Capital a/c

Capital a/c

| | $ | | $ |
|---|---|---|---|
| Drawings | 2,500 | Bal b/d | 1,000 |
| | | Bank | 1,500 |
| Bal c/d | 2,000 | Net Profit | |

c) Find Net Loss by the movement of Capital a/c

Capital a/c

| | $ | | $ |
|---|---|---|---|
| Drawings | 500 | Bal b/d | 1,000 |
| Net Loss | | Bank | 1,500 |
| Bal c/d | 1,000 | | |

*Classwork 1*

The cash balance as at 1 Jan 20X7 and 31 December 20X7 are $70 and $80 respectively. Total cash sales and cash purchase are $1,000 and $800 respectively. What is amount of cash drawings?

*Classwork 2*

Brenda Lau's bank account for the year 20X7 is as follows:

| | $ | | $ |
|---|---|---|---|
| Balance 1.1.20X7 | 1,890 | Cash withdrawn from bank | 5,400 |
| Receipts form debtors | 44,656 | Trade creditors | 31,695 |
| Loan from T. Ho | 2,000 | Rent | 2,750 |
| | | Rates | 1,316 |
| | | Drawings | 3,095 |
| | | Sundry expenses | 1,642 |

Records of cash paid were: sundry expenses $122, trade creditors $642. Cash Sales amounted to $698, cash drawings were $5,289.

The following information is also available:

| | 31.12.20X6($) | 31.12.20X7($) |
|---|---|---|
| Cash in hand | 48 | ? |

(a) What is the amount of Cash in the closing balance?.

## [03] Preparation for trading, profit and loss a/c from incomplete records

a) The steps in the preparation of final accounts from incomplete records

  (i) Drawing up an opening balance sheet / statement of affairs
- To find out the amount of opening capital

  (ii) Preparing summary cash and bank accounts
- To find out the amount of payment to specific item or;
- To find out the amount of receipt from specific item or;:
- To find out the amount of cash / bank drawings or;
- To find out the amount of closing balance

  (iii) Scheduling creditors
- To find out the amount of credit purchases or;
- To find out the amount of payment to creditors or;
- To find out the amount of closing creditors

  (iv) Scheduling debtors
- To find out the amount of credit sales or;
- To find out the amount of receipts from debtors or;

- To find out the amount of closing debtors

(v) To find out the amount of resalable stock and (or) closing stock

(vi) Making the year end adjustments (provision, reserves, prepayments)

(vii) Preparing the final accounts

b) The structure of Trading a/c

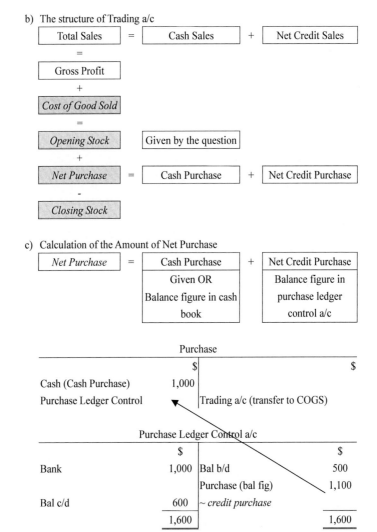

| Total Sales | = | Cash Sales | + | Net Credit Sales |

=

Gross Profit

+

*Cost of Good Sold*

=

| *Opening Stock* | | Given by the question |

+

| *Net Purchase* | = | Cash Purchase | + | Net Credit Purchase |

-

*Closing Stock*

c) Calculation of the Amount of Net Purchase

| Net Purchase | = | Cash Purchase | + | Net Credit Purchase |
|---|---|---|---|---|
| | | Given OR | | Balance figure in |
| | | Balance figure in cash | | purchase ledger |
| | | book | | control a/c |

Purchase

| | $ | | $ |
|---|---|---|---|
| Cash (Cash Purchase) | 1,000 | | |
| Purchase Ledger Control | | Trading a/c (transfer to COGS) | |

Purchase Ledger Control a/c

| | $ | | $ |
|---|---|---|---|
| Bank | 1,000 | Bal b/d | 500 |
| | | Purchase (bal fig) | 1,100 |
| Bal c/d | 600 | ~ credit purchase | |
| | 1,600 | | 1,600 |

d) Calculation of the Amount of Total Sales

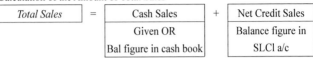

| Total Sales | = | Cash Sales | + | Net Credit Sales |
|---|---|---|---|---|
| | | Given OR | | Balance figure in |
| | | Bal figure in cash book | | SLCl a/c |

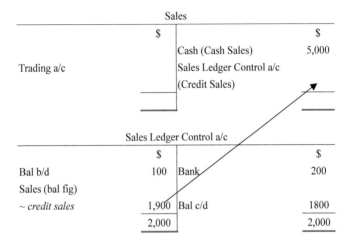

**Sales**

| | $ | | $ |
|---|---|---|---|
| | | Cash (Cash Sales) | 5,000 |
| Trading a/c | | Sales Ledger Control a/c | |
| | | (Credit Sales) | |

**Sales Ledger Control a/c**

| | $ | | $ |
|---|---|---|---|
| Bal b/d | 100 | Bank | 200 |
| Sales (bal fig) | | | |
| ~ credit sales | 1,900 | Bal c/d | 1800 |
| | 2,000 | | 2,000 |

*Classwork 3*

The following figures have been extracted from the records of Brenda who does not keep a full record of his transactions on the double entry system:

| 1 | April | 20X7 | Debtors | 2,760 |
|---|---|---|---|---|
| 1 | April | 20X7 | Creditors | 1,080 |
| 1 | April | 20X7 | Stock | 2,010 |
| 31 | March | 20X8 | Debtors | 3,090 |
| 31 | March | 20X8 | Creditors | 1,320 |
| 31 | March | 20X8 | Stock | 2,160 |

All goods were sold on credit and all purchases were made on credit. During the year ended 31 March 20X8, cash received from debtors amounted to $14,610 whereas cash paid to creditors amounted to $9,390.

(i) Calculate the amount of sales ad purchases for the year ended 31 March 20X8.

(ii) Draw up the trading account for the year ended 31 March 20X8.

**[04] Calculation of the amount of cash misappropriated**

The amount of the cash of organisation may be misappropriated.

| Sales Ledger Control a/c in order to calculate the "cash and cheque received" | Cash Book The amount of cash misappropriated |
|---|---|

**Sales Ledger Control a/c**

| | $ | | $ |
|---|---|---|---|
| Bal b/d | 200 | Bad Debts | 100 |
| Sales | 1,800 | Cash (bal fig) | |
| | | Bal c/d | 400 |
| | 2,000 | | 2,000 |

**Cash Book**

| | $ | | $ |
|---|---|---|---|
| Bal b/d | 500 | Cash Misappropriated (bal) | |
| Sales Ledger Control | | Bal c/d | 600 |

*Classwork 4*

1. Bank Deposit from Cash a/c is $972,000
2. The following amounts were paid out of cash: salary $15,000, sundry expense $24,000, drawings $20,000.
3. Cash Receipt from Debtor is $56,000
4. A refund of $12,000 cash from a supplier for goods returned has not been reordered.
5. Cash sales is $1,201,000

How much does cash misappropriated?

**[05]  Summary -** How to find the missing figure?

| Missing figures | Methods |
|---|---|
| Sales | <ul><li>Total debtors a/c / Sales Leger Control a/c</li><li>Trading a/c and margin</li></ul> |
| Purchases | <ul><li>Total creditors a/c / Purchase Ledger Control a/c</li><li>Trading a/c and markup</li></ul> |
| Stock Loss | <ul><li>Stock Turnover Ratio</li><li>Trading a/c</li></ul> |
| Capital | <ul><li>Statement of Affairs</li><li>Return on Capital Employed</li></ul> |
| Net Profit | <ul><li>Statement of Affairs</li><li>Return on Capital Employed</li><li>Different between opening and closing capital</li></ul> |
| Cash misappropriated | <ul><li>Cash Book</li></ul> |

= 本章完  End of Chapter =

[01]　補充 – 計數機特別使用法 (加數計算機如 CASIO JS-140)

1. SPECIAL FUNCTIONS OF CALCULATIOR

   1.1　Example 1　　A, B and C are in partnership sharing profits and losses in the ratio of 3: 2: 1 respectively.

                       Net Profit is $34,680

| CA | 3 | 4 | 6 | 8 | 0 |
|----|---|---|---|---|---|

Your calculator will display →　　　K　　Which represents constant.

A's share of Net Profit

3　=　　Your calculator will display →　　　| 17,340 |

B's share of Net Profit

| 2 | = | 　Your calculator will display →　　　| 11,560 |

C's share of Net Profit

| 1 | = | 　Your calculator will display →　　　| 5,780 |

Check Total

| GT |　　Your calculator will display →　　　| 34,680 |

1.2　Example 2

| Capital – A : $20,000 |
| Capital – B : $36,800 |
| Capital – C : $19,700 |

Interest on Capital is 12% p.a.

| CA | . | 1 | 2 | X | X |
|----|---|---|---|---|---|

Your calculator will display →　　　K　　which represents constant.

A's Interest on Capital

| 2 | 0 | 0 | 0 | 0 | 0 | = |
|---|---|---|---|---|---|---|

Your calculator will display →　　　| 2,400 |

B's Interest on Capital

| 3 | 6 | 8 | 0 | 0 | = |

Your calculator will display → | 4,416 |

C's Interest on Capital

| 1 | 9 | 7 | 0 | 0 | = |

Your calculator will display → | 2,364 |

Total Interest on Capital

| GT | Your calculator will display → | 9,180 |

[02] 出題新方向 – 混合式題目

| Provision for Depreciation | + | Capital & Revenue Expenditure |

| Accounting Ratio | + | Comments |

| Cash Book | + | Bank Reconciliation Statement |

| Partners Retirement | + | Partners Admission |

| Accounting Ratios | + | Incomplete Records |

| Correction of Errors | + | Control a/c |

| Stock Valuation | + | Stock Loss |
|---|---|---|

| Manufacturing a/c | + | Issue of Shares & Final Accounts |
|---|---|---|

| Accounting Ratios | + | Cost of Stock Destroyed |
|---|---|---|

[03] 各課題的問題範例

| 課題 | | 問與答 |
|---|---|---|
| 原始紀錄 | 問 | State the differences between trade discounts and cash discounts. |
| | 答 | 1. *計算方法:* *Trade discounts directly deduct from the list price of the goods or services before trading.*<br>2. *入帳處理:* *It usually records in the sales day book or purchase day book*<br>3. *目的:* *It is offered when the large quantity of goods sold or maintaining the relationship with customers*<br><br>1. *計算方法:* *Cash discounts usually record in the (three column) cash book.*<br>2. *目的:* *It aims to make payment early*<br>3. *入帳方法:* *It will post to discount allowed a/c or discount received a/c.* |
| 折舊準備 | 問 | What are the causes of depreciation? |
| | 答 | *Physical deterioration, amortization, damage in unusual, obsolescence* |
| 呆帳準備 | 問 | What is the difference between bad debts and provision for doubtful debts? |
| | 答 | *Refer the text book chapter 2.* |

| | | |
|---|---|---|
| 統制帳 | 問 | State briefly any two advantages of maintaining control accounts |
| | 答 | *Bookkeeping errors can be located quickly* |
| | | *Total figures for creditors and debtors can be quickly produced* |
| | | *Separation of duties reduces the possibly of fraud* 欺詐. |
| | | *The control a/c provides an automatic check of the postings in the ledger.* |
| 錯誤更正 | 問 | List all the types of errors that not affecting trial balance agreement. |
| | 答 | *1) Errors of Commission 2) Errors of Principles 3) Errors of Omission 4) Errors of Original Entry 5) Compensating Errors 6) Complete reversal of entries* |
| 現金簿 | 問 | What is dishonoured cheque? State the causes. |
| | 答 | *It means that the cheque cannot be presented because:* |
| | | • *Insufficient fund* |
| | | • *Wrong in the drawer's signature* |
| | | • *Wrong in the date of the cheque* |
| 存貨價值 | 問 | State three methods for calculating the value of stock. What means by NRV? |
| | 答 | *FIFO, LIFO, WAVO* |
| | | *NRV: Saleable value – expenses incurred before selling* |
| 合夥經營 | 問 | What elements should be included in a partnership agreement? |
| | 答 | *Capital Contribution, Profit/Sharing Ratio, Interest on Capital, Interest on Drawings, Partner Salaries, Arrangement for the admission, Arrangement for a partner retires* |
| | 問 | What happens if there is no partnership agreement? |
| | 答 | *Profit and Loss share equally, No interest on capital, No interest on drawings, No partners salaries, Any partner withdraw or join need to be agreed with all partners* |
| | 問 | In what situations that partnership goodwill should be revalued? |

| | 答 | *Change in Profit / Loss sharing ratio, admission of new partners, retirement of existing partners, sales of the partnerships* |
|---|---|---|
| 會計比率與 分析 | 問 | List three ratios that can reflect the profitability for a company. |
| | 答 | *Gross Profit Ratio, Net Profit Ratio, Return on Capital Employed* |

[04]熱門課題分析

(a) Correction of Errors and Suspense a/c
這課題因出題範圍極度廣泛，故比較困難，在試卷的出題率謹次於夾算帳，考生須特別留意。

答題步驟
(I) 首先詳細閱讀題目，小心完成 Journal 關於錯誤更正的部份
(II) 按已做的錯誤更正，編制 Suspense a/c[注意：只有部份錯誤涉及 Suspense a/c]
(III) 按照已完成的 Journal，修正新的 Net Profit / Gross Profit

> Example
> Return Outwards of $200 to Richard had been recorded only in the personal account ("ONLY" is the key word)

處理步驟：
(I) 先列出正確的入帳方法，然後再列出題目的錯誤入帳法，最後提出改正方法

| | Dr | Cr |
|---|---|---|
| 正確 | Creditor - Richard | Return Outwards |
| 錯誤 | Creditor - Richard | --- |
| *改正* | *Suspense* | *Return Outwards* |

(II) 根據步驟(I)，編制 Suspense a/c

Suspense

| | $ | | $ |
|---|---|---|---|
| Return Outwards | 200 | | |

(III) Return Outwards 會影響 Gross Profit 及 Net Profit，Net Profit 調整如下：

Brenda Company
Statement of adjusted Net Profit as at …….

|  | $ |
|---|---|
| Net Profit before adjustment | xxx |
| Add: Understate the Return Outwards | 200 |
| Net Profit after adjustment | xxx |

(b)  Accounting Ratios
此課題有兩種兩核方法：
1.   利用題目的資料直接計算比率，然後加以評論分析 (比較容易)
2.   推理式試題，利用題目中的 Ratios 及零碎資料完成不完整的帳戶，例如： Trading, Profit & Loss Appropriation a/c 及 Balance Sheet

直算式答題技巧：
應分別列出方程式，再列出計算過程，最後才寫出答應。

| 方程式 | | | 計算過程 | 2007 年 | 2008 年 |
|---|---|---|---|---|---|
| ROCE | = | Net Profit / Capital Employed | | | |
| … | | | | | |
| …. | | | | | |
| … | | | | | |

推理式答題技巧
作答推理式題目，應細題目中的零碎資料及比率，然後逐一將遺失數字尋找出來，那當然，尋找的次序是不能有誤，而要準確地找出推理模式，相信要靠經驗的累積了。

Example:

Prepare the trading and profit and loss account and calculate the gross profit.
Margin = 20%    Sales = $20,000    Return Outwards = $2,000
Stock Turnover rate = 4 times    Opening stock = $30,000

Trading, Profit & Loss a/c for the year ended 31 December 20X7

|  |  |  | $ |  | $ |  | $ |
|---|---|---|---|---|---|---|---|
| Sales |  |  |  |  |  |  | 200,000 |
| Less: | Cost of Good Sold |  |  |  |  |  |  |
|  | Opening Stock |  |  |  | 30,000 |  |  |
|  | Purchases | [5] | 182,000 |  |  |  |  |
|  | Return Outwards |  | (2,000) | [4] | 180,000 |  |  |
|  | Closing Stock |  |  | [3] | (50,000) | [2] | 160,000 |
| Gross Profit |  |  |  |  |  | [1] | 40,000 |

STEP

[1]  Gross Profit = Sales x margin = $200,000 x 20%

[2]  COGS = Sales – Gross Profit = 200,000 – 40,000 = 160,000

[3]

$$\text{Stock Turnover Rate} = \frac{\text{Cost of Good Sold}}{\text{Average Stock}}$$

$$4 = \frac{160,000}{(30,000 + CS) \times 0.5}$$

$$4 = \frac{160,000}{15,000 + 0.5CS}$$

| 60,000 + 2CS | = | 160,000 |
|---|---|---|
| 2CS | = | 100,000 |
| CS | = | 50,000 |

[4]
| OS + Net Purchase – CS | = | Cost of Good Sold |
|---|---|---|
| 30,000 + Net Purchase – 50,000 | = | 160,000 |
| Net Purchase – 20,000 | = | 160,000 |
| Net Purchase | = | 180,000 |

[5]
| Total Purchase – Return Outwards | = | Net Purchase |
|---|---|---|
| Total Purchase – 2,000 | = | 180,000 |
| Total Purchase | = | 182,000 |

(c)  Provision for Depreciation
- 首要注意該資產所使用的折舊方法，直線法／餘額遞減法
- 注意該資產的購買日期及年結日期，可能不足一年，要清楚計算月份，建議以數手指方法最可靠
- 另外，一些必要開支，如運輸費 Transportation，應包括於固定資產帳內。計算 Depreciation 時必須包括該開支
- Disposal 的步驟及日期亦應特別注意

(d) Control a/c

要於 Control a/c 中取得較高分數,考生必須知道統制帳中各金額的來源,
詳情可見後表。

| Sales Ledger Control a/c | | Purchase Ledger Control a/c | |
|---|---|---|---|
| **Information** | Sources | **Information** | Sources |
| **Opening Balance** | List of all debtor accounts drawn up at the start of the period | **Opening Balance** | List of all creditor accounts drawn up at the start of the period |
| **Credit Sales** | Sales day book | **Credit Purchase** | Purchase day book |
| **Return Inwards** | Return Inwards Day Book | **Return Outwards** | Return Outwards Day Book |
| **Cheque received** | Bank a/c | **Cheques paid** | Bank |
| **Cash received** | Cash a/c | **Cash paid** | Cash a/c |
| **Discount Allowed** | Discount column in Cash Book | **Discount Received** | Discount column in Cash Book |
| **Bad Debts** | Journal | --- | --- |
| **Closing Balance** | List of all debtor accounts drawn up at the end of the period | **Opening Balance** | List of all creditor accounts drawn up at the end of the period |

[11]  後記

- 充份準備才能使你的壓力下降!
- Sosir2006@hotmail.com 全天候支援